Smith's
Fundamentals of Motorsport Engineering

Josh Smith

Nelson Thornes

Text © Josh Smith 2013
Original illustrations © Nelson Thornes Ltd 2013

The right of Josh Smith to be identified as author of this work has been asserted by him in accordance with the Copyright, Designs and Patents Act 1988.

All rights reserved. No part of this publication may be reproduced or transmitted in any form or by any means, electronic or mechanical, including photocopy, recording or any information storage and retrieval system, without permission in writing from the publisher or under licence from the Copyright Licensing Agency Limited, of Saffron House, 6–10 Kirby Street, London, EC1N 8TS.

Any person who commits any unauthorised act in relation to this publication may be liable to criminal prosecution and civil claims for damages.

Published in 2013 by:
Nelson Thornes Ltd
Delta Place
27 Bath Road
CHELTENHAM
GL53 7TH
United Kingdom

13 14 15 16 17 / 10 9 8 7 6 5 4 3 2 1

A catalogue record for this book is available from the British Library

ISBN 978 1 4085 1808 3

Cover photograph: all-free-photos.com

Page make-up by GreenGate Publishing Services, Tonbridge, Kent

Illustrations by GreenGate Publishing Services and Barking Dog Art

Printed in China

Author's acknowledgements

For their support and help over the past six months of writing this book, I have many people to thank for various different reasons.

Firstly I need to thank Anna (my girlfriend), for supporting me, giving me the motivation to keep going from start to finish and for being patient during times when I was rather stressed.

All of the technical reviewers who have helped with guidance for each chapter have been brilliant and they include; Mark Bailey of Mark Bailey Racing who had the biggest part to play of all reviewers, along with Frank Anderson of Anderson Racing Engines, Hugh Mattos of Explorer Marine and who is a lecturer at Bridgwater College and Tim Mason of Tim Mason Motorsport Electronics.

Dave Swan, who is a former student at the college and now a lecturer, was also great in providing the 'student view' of the book and looked through all of the chapters along with Joel Vermiglio, during the proofing phase.

I must also thank all of my friends and family who have helped me get to a stage where I am capable of writing and completing this book, which is mainly down to the drive from my parents (Darcy and Helen Smith) from an early age.

And of course everyone at Nelson Thornes who gave me the opportunity to write the book and helped all the way through including Helen Broadfield, Phil Gallagher, Jenni Johns, Rachel Howells, and Rebecca King.

Thank you all again.

Bridgwater College race car, student team and Josh (in car)

We would like to thank these companies for their valuable contributions:

Contents

Preface vi

Introduction vii

1 Engine 1
1.1 Two-stroke and four-stroke cycles and the basic principles 2
1.2 Inlet system 5
1.3 Carburettors and fuel injection 7
1.4 Ignition system 15
1.5 Cylinder head 16
1.6 Exhaust system 23
1.7 The bottom end 26
1.8 Lubrication and cooling 29
1.9 Forced induction 34
1.10 Dynos and tuning 36

2 Transmission 39
2.1 Flywheel 40
2.2 Clutch 42
2.3 Gearbox 50
2.4 Drive shaft and propshaft 62
2.5 Differential 64

3 Chassis 75
3.1 Tyres 76
3.2 Suspension 82
3.3 Steering 102
3.4 Brakes 105

4 Aerodynamics 125
4.1 Principles 126
4.2 Body shape 132
4.3 Wings 134
4.4 Underfloor and diffusers 138
4.5 Additional parts 140

5 Electrical 145
5.1 Sensing 146
5.2 Wiring harnesses 157
5.3 Batteries and components 158
5.4 Driver interface 162
5.5 Controller area network (CAN) 166

6 Data logging 167
6.1 Data logging systems 168
6.2 Telemetry 170
6.3 What to monitor – driver/chassis/engine 171
6.4 How to analyse the data 171

7 Basic engineering and preparation 185
7.1 Preparing a car 186
7.2 Checklists 195

8 Event and set-up 199
8.1 What happens at an event 200
8.2 Getting involved in motorsport 206
8.3 Set-up 208

Glosssary 218

Index 224

Acknowledgements 232

Preface

Motorsport has been part of my life from a very young age, be it preparing and supporting cars or racing myself. My key area of interest tends to be circuit racing and, in particular, sports prototypes and single-seaters. These cars interest me because their sole design objective is to be able to lap a circuit as fast as possible; the rest of the design criteria have to fit around this. These lightweight cars, fitted with aerodynamic aids, can generally outperform the most powerful saloon cars and super cars.

I have written this book to provide the reader with a good level of general knowledge of the subject of motorsport. I have tried to include the most up-to-date content in a way that is accessible for students, lecturers, enthusiasts and professionals within the industry.

Figure 0.1 The author after a race at Castle Combe

Introduction

Motorsport is a niche industry and one that demands constant development in all areas in order for teams to keep up with each other. It is one that speaks the same language across the world and one that has the same aim, no matter what discipline or level of the sport you are involved in. 'If you stand still, you go backwards' could not be a truer statement.

Motorsport has evolved hugely since its inception in terms of components and car shapes, however, the principle items have largely remained the same for the last 10–20 years or more (although we have seen an improvement in the accessibility of items such as composite and other exotic materials).

The first documented and organised race was way back in the late 1800s in France, with the first purpose-built circuit being the Milwaukee Mile, USA in 1903. In the UK, the first purpose-built venue was Brooklands in Surrey in 1907. Most race circuits in the UK have evolved from airfields, including Silverstone, Thruxton and Castle Combe, and the UK boasts a great variance in circuits.

Since then, motorsport has flourished and has many different levels and categories. Among these are circuit racing, rallying, drag racing, sprints and hill climbs, autotests, rallycross and trials, but there are many more.

Under the circuit racing banner there are many different types of racing car, such as Formula cars, touring cars, sports cars, kit cars, saloon cars and historics – definitions for all can be found in the MSA Blue Book. The MSA Blue Book outlines the rules and regulations for all forms of motorsport that come under the Motor Sports Association (MSA). There are also One-make series races, in which only one manufacturer of car is raced, which generally provides a fairer level of racing.

The UK has a very good racing heritage and has had great success in the manufacture of race cars, such as Aston Martin, Brabham, Cooper, Caterham, Mallock, Lotus and Radical, to name but a few. Not to mention the fact that nearly all Formula 1 teams are based in the UK.

Figure 0.2 Sports racers at Silverstone

The industry itself is very demanding and to work in it requires specific skill sets for each area, be it for the team, consultancy company, parts manufacturer or in a freelance role.

This book aims to provide you with an insight into all the main areas of motorsport without needing to buy individual books for each part of this highly technical engineering industry. While this may not give you the depth of detail as other, more specific books, it will give you a good level of understanding of each area.

The book is written to appeal to the enthusiast but is also carefully written to suit those studying at a minimum of Level 3. To this end, there is an assumption that you have a basic understanding of the motorcar. If this is not the case, I suggest you buy a copy of *Hillier's Fundamentals of Motor Vehicle Technology* (see the back cover of this book).

My own specialism is circuit racing, with a specific focus on sports racing cars (or sports prototypes) and Formula cars (also known as single-seaters). As a result, this book is focused around this area. To cover all areas of motorsport for all types of car and events would either result in content that would be too brief or a book that would be too big!

The book includes both theory and practical elements, with a range of diagrams and photos to support the technical content and help put the theory into context. Applied calculations will be used where possible to aid understanding and application. I hope you enjoy the book and find it a useful addition to your involvement in the motorsport industry.

Figure 0.3 A Formula 3 car

CHAPTER 1

Engine

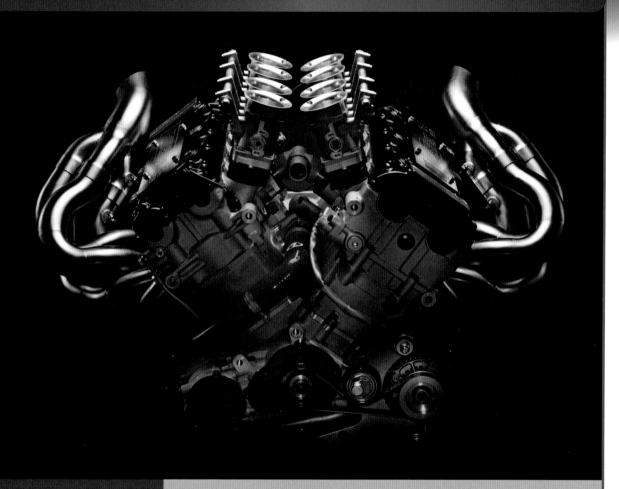

This chapter will outline:

- **Two-stroke and four-stroke cycles and the basic principles**
- **Inlet system**
- **Carburettors and fuel injection**
- **Ignition system**
- **Cylinder head**
- **Exhaust system**
- **The bottom end**
- **Lubrication and cooling**
- **Forced induction**
- **Dynos and tuning**

1.1 Two-stroke and four-stroke cycles and the basic principles

We will quickly cover the basic operation of the engine to remind the reader of just how the engine works before jumping into the deep end of engine technology. Engines are an easy way to extract better lap time from a race car and it is often the way that the better-financed teams find an advantage over others, by spending thousands to gain extra power and torque within the constraints of series regulations.

Throughout the chapter we will assume the use of a four cylinder engine.

1.1.1 Two-stroke engines

The two-stroke engine is still used in some motorbike racing categories and, as shown in Figure 1.1, it is a relatively simple system with minimal moving parts. It has a power stroke every two strokes and the cycle is complete in one revolution of the crankshaft.

On the upstroke, the piston moves upwards and creates a low-pressure area underneath it. The inlet port is then exposed and allows the air and fuel mixture to be drawn into the crankcase (remembering that this will be mixed with a ratio of oil to lubricate the system). At the end of the upstroke, both ports are closed off and the fuel mixture is compressed just before the spark plug fires prior to the piston reaching top dead centre (TDC).

On the downstroke, the combustion process forces the piston back down the cylinder bore. As it moves down, it passes and then unveils the exhaust port, allowing the burnt gases to escape.

As the piston then clears the transfer port at the top, the pressure underneath the piston is then expelled by allowing the fresh air and fuel mixture into the top of the cylinder, which also helps to force out the remaining burnt gases.

1.1.2 Four-stroke engines

The four-stroke engine (Figure 1.2) has had continual development over the years and it is now possible to find figures of nearly 300 bhp from naturally aspirated 2.0l production engines, such as the popular Vauxhall XE engine, along with the Ford Duratec and Honda VTEC engines. Motorcycle engines are still pushing the boundaries year on year, with each manufacturer striving to produce the top 1000 cc engine – current engines producing 190–200 bhp in standard form.

The four-stroke engine has a power stroke every four-strokes and the cycle is complete in two revolutions of the crankshaft. The four strokes are as follows:

- Inlet stroke – the piston travels down the bore as the inlet valve opens and the air/fuel mixture enters the cylinder.

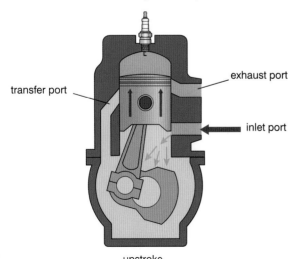

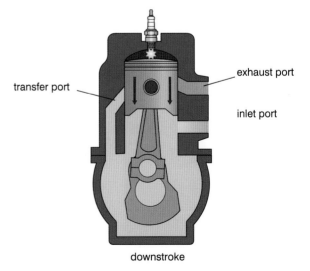

Figure 1.1 Two-stroke cycle

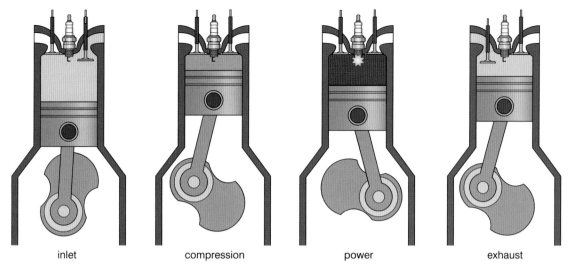

inlet　　　　　　compression　　　　　　power　　　　　　exhaust

Figure 1.2 Four-stroke cycle

- Compression stroke – with both valves shut, the piston moves up towards TDC to compress the air/fuel mixture together. As the piston reaches near TDC, the spark plug ignites the mixture. This firing of the spark plug to ignite the mixture and its timing were directly controlled by a distributor but these have been replaced by an electronic control unit (ECU) and coil pack, which takes inputs from a variety of sensors from the engine to ensure timing and control is optimised.
- Power stroke – as the rapid burning process of the mixture commences, the pressure created forces the piston back down the bore.
- Exhaust stroke – with the piston now moving back up the bore, the exhaust valve is opened to allow the burnt mixture to escape via the exhaust port and then out through the exhaust.

1.1.3 Basic principles

Engine tuning is a fine art and some basic principles can be assumed in order to maximise the objectives of tuning an engine to suit a set of race series regulations.

There are a few key areas to consider with the engine, which will be broken down further in this chapter, but using the basic theory that the internal combustion engine is just an air pump, we can assume that:

- the higher the quality of the air/fuel mixture that enters and exits the engine the better – this includes high-density air, atomised fuel, full cylinders and a scavenging exhaust system
- with correct ignition timing of the air/fuel mixture, the rate of burn is controlled to release the energy in the correct manner
- a complete system must be built, with components that all work in harmony with each other.

High air density is hugely important in a race engine and there are a few things that affect it. Heat is the most common of these. As temperature increases, the air expands and the oxygen becomes more spaced out, resulting in a lesser charge in the cylinders – you can lose roughly 1 per cent of engine horsepower (hp) for every 7 °C increase in air temperature. Altitude also has an effect on air density. Air density at sea level is considered to be 14.7 psi, and at 1000 ft above sea level the air pressure is 14.2 psi – a loss of 0.5 psi – so horsepower decreases by approximately 3 per cent for every 1000 ft. Humidity does not have as large an effect on air density, however, high humidity causes the air to effectively weigh more. The oxygen content of this heavier air is not higher, only the water level is. So, ideally, we need to be racing with cold, dry air at low altitudes. Now that may be difficult, but there are some things that we can do to help our situation which will be discussed.

The fuel entering the engine must be of a high quality and be atomised correctly to allow it to mix effectively with the air, creating a more controlled, stronger and more complete burn in the cylinder.

Having good flow in and out of the cylinders will allow volumetric efficiency to increase, giving the engine better capability to fill the cylinders.

The quality of this mixture (air and atomised fuel) and airflow in and out of the cylinders, coupled with a correct exhaust design, will allow the cylinders to be scavenged of all burnt gases when exhaust valves open. Exhaust lengths and diameters can then also be adjusted, which allows the **torque** and **power** curves to be tuned.

When deciding on an engine package, it is vital to get the right balance so that all the components internal and external to the engine are matched in terms of performance level. For example, large throttle bodies and a large bore exhaust manifold will be of no use to a cylinder head that cannot let enough air in and out due to its small valve sizes. Similarly, selecting a full race camshaft that produces peak power at 8500 rpm is of no use if the bottom end of the engine is only sufficient to rev to 7500 rpm.

Engines can be specified in various states of tune:

- Standard – no modifications and as produced by the manufacturer.
- Fast road – general top-end tuning of the engine that still allows low revs per minute (rpm) driveability. This could include headwork, camshafts and a different inlet and exhaust system but to a mildly tuned state.
- Semi-race – this is for the club racer who has the ability to modify gear ratios, for instance, to cope with a smaller power band higher up in the rev range, although it still maintains some standard manufacturer components, such as crankshaft and conrods.
- Full race – an all-out competition engine with no regard for cost, built to produce maximum torque and power with all engine internals replaced to cope with the increase in stress due to high compression, lightweight parts and very low tolerances. This engine is likely to have a relatively narrow power band and reduced driveability at low to mid rpm. (See Figure 1.3.)

Figure 1.3 Full race engine

> **Torque:** a measure of the force that is applied on a lever, multiplied by the distance to the rotation point. Torque = Force × Distance from pivot

> **Power:** how fast work can be done in a given time. It is a function of torque and engine rpm. Power (hp) = (Engine rpm × Torque (lb/ft)) ÷ 5252

Most small capacity bike engines will tend to rev very high so that they can overcome their small torque figure with high revs to produce power. Other ways to affect the engine's power band can be forced induction, variable valve timing and of course engine displacement.

1.2 Inlet system

The air inlet system is one of the most commonly modified areas, closely followed by the exhaust system, due to the ease of its replacement. You must then also make sure that the cylinder head can breathe to the same extent as the inlet and exhaust.

1.2.1 A modified system

When utilising the standard intake system in terms of a single throttle body and original intake trunking, it is important to remove any restrictions of airflow, such as resonator boxes, replacing vane/flap-type airflow meters with hot wire meters – removing the mesh also (or by using a manifold absolute pressure (MAP) sensor instead) and replacing the filter with a higher flow type. The flaps can also stop dirt and grit from entering and damaging any engine components. As stated earlier, keeping the air as cold as possible is important for air density, so shielding the inlet system from any heat source is vitally important.

A great addition is a K&N cotton gauze filter, as it provides good flow and filtration properties. It is easily cleaned and oiled. Running without a filter with the aim of increasing flow can cause a problem, as the additional wear of internal engine components through the intake of particles of grit, for instance, can start to hamper valve and piston ring sealing properties, incurring costs.

When considering further modifying the air inlet system, ram air should be considered. Motorcycle designers and single-seater constructors spend a lot of time designing the airbox to help increase air pressure and allow the engine to have a free stream of clean and cool air. Simply sticking an air scoop out of the engine cover will not suffice. Other considerations need to be taken into account, such as air turbulence, air scoop inlet design and also ensuring that each individual inlet tract has a relatively equal supply of air pressure across the airbox. This equality of air pressure will allow each cylinder to work at its maximum capability. It is possible to see around a 2 per cent rise in air pressure when travelling at 120 mph. You should then expect to see a 2 per cent rise in horsepower, unless turbulence is affecting the mixture or the carburettors are being pressurised and may be making the mixture richer. Wind tunnel testing and computational fluid dynamics (CFD) testing are great ways to look at airstreams and the effect of internal and external flow that an air scoop may produce.

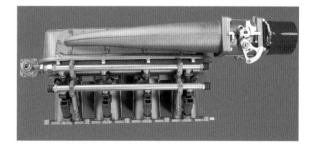

Figure 1.4 Inlet systems

To summarise, the inlet system, up until the throttle body/carburettor, is vital and can affect the rest of the engine as this is the first part of the engine that can be used to help tune the system. It can increase airflow and stop dirt and grit from entering and damaging any engine components. To work at its best, it must be free of restriction, shielded from any heat source and exposed to a fresh stream of air.

1.2.2 Race system set-up

Assuming a race system where the single throttle body has been replaced with either individual throttle bodies (ITBs) or a pair of twin Weber carburettors, the inlet manifold is a key area to consider, not only for its length and bore but also so that it is matched to the inlet ports and the throttle bodies/carburettors.

Pulse tuning the inlet runner lengths can really affect the power curve based upon the length and diameter of the inlet runner. The inlet runner length is considered to be the measurement from the valve seat all the way back to the trumpet face on the end of the throttle bodies/carburettors.

The length and diameter are very important and can have the following effects:

- The diameter fixes the rpm at which the engine makes its best horsepower. A wide diameter runner will allow high flow at high rpm and, therefore, power, but at low rpm the power will be down due to low air speeds.
- The runner length will pivot the power curve around the peak horsepower mark and so a long runner will give more power at low rpm, while a short runner will give more power at higher rpm.
- Variable length intake runners, controlled by the ECU, and triggered actuators can be used to try and gain the best of both worlds and increase average horsepower across the whole rev range.

In practice, however, pulse tuning to a specific rpm range is dependent on internal components' properties, such as the con rods and crankshaft, which will give the ability to extend to the higher rev ranges when upgraded.

To show the effect of how tuned lengths can change the peak tuned engine speed, we can use the following equation:

$$TS = 642 \times 350.52 \times \sqrt{\frac{A}{L \times V_d}} \times \sqrt{\frac{r-1}{r+1}}$$

TS = Tuned speed (rpm)

A = Cross-sectional area of runner (cm^2)

L = Length of runner (cm)

V_d = Cylinder size (cc)

r = Compression ratio

Applying some numbers to this:

A = 16, L = 30, V_d = 500 and r = 12

$$TS = 642 \times 350.52 \times \sqrt{\frac{16}{30 \times 500}} \times \sqrt{\frac{12-1}{12+1}}$$

TS = 22,5033.84 × 0.03266 × 0.919866 = 6761 (to nearest whole number)

This gives a TS of 6761 rpm, and by lengthening L to 40 cm gives us 5855 rpm.

Although exhaust tuning and the cam type can also affect this peak tuned rpm, they are not taken into account here.

The goal of intake tract tuning is to control or alter the rpm at which peak volumetric efficiency of the engine will occur.

Periodic opening and closing of the inlet valves creates a fluctuating column of kinetic energy. This kinetic energy can be used to effectively 'ram' the induction charge into the cylinder over a limited engine speed range. This speed range is determined by the inlet tract's length and diameter.

The pressure drop caused by the rapidly accelerating piston and the open inlet valve causes the column of air in the intake tract to rapidly accelerate towards the cylinder. This fast-moving or 'charged' air is then brought abruptly to a halt by the closing inlet valve. This converts the kinetic energy in the air into pressure in the closed tract.

The area of high pressure now immediately behind the inlet valve 'reflects' back along the intake tract at the speed of sound until it meets the atmospheric pressure at the opening. At this point, it suddenly becomes less dense and so creates an area of low pressure. The surrounding air rapidly fills this low pressure and creates a positive pressure wave, which travels back to the inlet valve port.

If this reflected wave is timed correctly with the opening of the inlet valve, it is possible to ram the charge into the cylinder, thus increasing the volumetric efficiency.

An inlet manifold with individual inlet runners allows this parameter to be fine-tuned. Trumpet lengths can be used as fine-tuning tools, dependent upon under-bonnet space (see Figure 1.5).

Figure 1.5 Various lengths of inlet trumpets

1.3 Carburettors and fuel injection

There are pros and cons to using both systems but fuel-injection systems are becoming more affordable, along with all of the electrical components required. Carburettors have the key benefit of being simple, comparably cheap, reliable and potentially lighter as a complete system – the disadvantage of the system being that the tunability across the whole rev range is restricted and can be a compromise, dependent upon application. A fuel-injection system with throttle bodies, a distributorless ignition system and an ECU to control fuelling allows for much finer control of fuel-injection timing and duration across the whole rev range. It does bring on a few potential problems of its own in that a conversion would require a high-pressure fuel pump, different fuel lines and the addition of a crankshaft position sensor and a different ECU, which is costly and could be a source of reliability issues if not installed correctly.

1.3.1 Carburettors

In most cases, carburettors can provide nearly as much peak power as an engine using fuel injection, but they can fall down in the driveability stakes when looking at a full race engine with big cams and a large valve head.

Weber DCOE carburettors (Figure 1.6) are the most common carburettors used in racing. Their main benefit is that they are adjustable and can be modified to suit a variety of different engines in different states of tune.

A Weber carburettor is sized in terms of its physical inner bore. The DCOE carburettor, which is a sidedraught system, can come in the following sizes: 40mm, 42mm, 45mm, 48mm and 50mm. Then there is the venturi (or choke), which puts a sleeve inside of the carburettor to control airflow and speed. The venturi provides a pressure differential, which effectively triggers the main jet to discharge fuel – the venturi can be interchangeable. These affect the peak power of the engine and also the low- to mid-range driveability. If you are only searching for peak power and fit large venturis, you will find that fuel atomisation is only good at high rpm and so the engine will run poorly at low rpm. You will also find that, due to the range of carburettor and venturi sizes, there is a lot of overlap between the ranges of configurations available.

An example for a 2.0 litre engine running different configurations of twin DCOE Weber and venturi sizes can be seen in Table 1.1.

8 Engine

Figure 1.6 Weber carburettors

State of tune	Size of carburettor		
	2 × 45 DCOE	2 × 48 DCOE	2 × 50 DCOE
Fast road	34	–	–
Semi-race	36	–	–
Full race 7500 rpm	38	38	–
Full race 8500 rpm	40	42	42
Full race 9500 rpm	–	45	45

Table 1.1 Configurations of twin DCOE Weber and venturi carburettors for a 2.0 litre engine

Inside the carburettor are three key sub-systems:

- High-speed – main jet, air corrector (air bleed) and emulsion tube. These all come in a range of sizes and are a single assembly when fitted into the Weber carburettor. The main jet will be stamped with a number, for example '180', which means a 1.80 mm bore. The main jet is pressed into the bottom of the emulsion tube with the air corrector pressed into the top of the tube. The air corrector uses the same numbering code as the main jet but is often found to be slightly larger than the main jet that is in use. By increasing the size of the air corrector, you will weaken the mixture at higher rpm. Changing the main jet will alter the mixture equally across the rev range. The emulsion tube emulsifies the fuel previously metered by the main jet and affects the performance at small throttle opening angles and during acceleration. The emulsion tube can have different diameters along with different sizes, positions and numbers of holes – it controls the amount of fuel that is drawn into the airstream and is used to tune the mid range of the engine.
- Accelerator – the accelerator pump jet has the main function of applying an amount of fuel to help with acceleration and avoid getting a flat spot as, when the throttle is floored, the air will accelerate quicker than the fuel. This is because, when the throttle plate is opened quickly, air is sucked through quicker than the fuel due to density differences. It also bleeds additional fuel at high speeds to help with mixture richness and can be used to assist with cold starting by pumping the throttle pedal a couple of times before attempting to start the engine.

- Idle – the idle jet controls the fuel mixture at low speeds and when the engine is idle. This is because the main jet does not have enough air being drawn through the venturi to allow it to operate, due to the throttle plate being opened at only a very small angle. The idle jet has a fuel hole and an air bleed hole.

It is important to mount carburettors using some form of anti-vibration system, as mounting them solidly will cause the carburettors to vibrate and could cause fuel frothing, flooding and inaccurate metering of the fuel. Good engine builders will often use lock wire and silicone to ensure that specific parts of the carburettors (e.g. the venturi grub screws) do not come loose, as this would result in the engine running poorly.

Another simple but highly important part of the carburettor set-up to get correct is the throttle linkage. This must be sturdy and not flexible and, regardless of how many carburettors are used, the linkage must open all throttle plates at exactly the same time from closed to wide-open throttle (WOT) in order to allow the engine to run smoothly. Vacuum gauges can be used to synchronise each inlet system on the carburettor. An alternative cheap and basic solution is to use a rubber pipe and the human ear. By placing one end of the pipe close to the throttle flap and the other by your ear, you will hear a distinct hissing noise as air is drawn into the carburettor. From this noise, you can then match the rest of the carburettors to that same noise (equalise them all) by adjusting each carburettor's throttle flap position on idle. This should synchronise the carburettors or throttle bodies to a decent standard. Once set up on a dyno, it is a good idea to test it on the track as well. The engine should be able to accelerate both gradually and when the throttle is flattened without any hesitation and with no black smoke emitting from the exhaust, which shows signs of richness.

1.3.2 Fuel injection

The modern equivalent of carburettors is electronic fuel injection, which allows greater control and precision across the whole rev and load range in comparison to carburettors. Mapping of the fuel injectors allows the pulse width of the injectors to be controlled by the ECU, which takes inputs from various sensors, such as crank position, throttle position and rpm, to determine the precise amount of fuel that needs to be injected. The ECU will use data that is stored in its read-only memory (ROM) look-up table to compare sensor inputs to quickly trigger the fuel injectors for the specific pulse width and switch the fuel pump to control pressure.

The basic principles of throttle body sizing is very similar to carburettors: a smaller throttle plate will assist in mixing the air and fuel better across the whole rev range, giving greater throttle response, better low- and mid-range power, but restricting maximum power; a larger throttle plate gives an increased peak and high rpm power but can reduce low- and mid-range performance. Fuel injection will always give better responsiveness and driveability due to its high level of control and monitoring.

Figure 1.7 Throttle bodies

Race engines will use individual throttle bodies (ITBs), which allow each cylinder to have its own air and fuel supply while also doing away with the airflow meter, which can restrict flow. Engine speed and a throttle position sensor are used to determine airflow.

Other sensors that may be required for this system:

- Coolant temperature sensor – this allows the ECU to detect when the engine is cold and can increase the mixture richness until it reaches a set temperature level where the engine will run smoothly on a normal mixture.
- Air temperature sensor – this can also be used to alter fuelling, as air density changes with temperature. As the air temperature increases, the fuelling is reduced to ensure the correct ratio of air to fuel is maintained.
- Fuel temperature sensor – for a given volume of petrol, there are more molecules of fuel when it is cold compared to when it is hot, so the mixture will slightly weaken as the fuel heats up – this could damage the engine unless the ECU compensates for this.
- Lambda or oxygen sensor – this can determine the oxygen content in the exhaust, which then shows whether the mixture is rich (low oxygen content) or lean (high oxygen content) and can allow the ECU to alter fuelling. Most engine mappers will keep the engine at 12.5–13:1 as an air/fuel ratio, but this will vary across the rev range to find maximum power from low through to high rpm.
- Exhaust gas temperature sensor – this uses a temperature sensor in each exhaust primary. Every engine will run at a different temperature (e.g. 800°C) but the ECU (dependent upon its ability) will be able to ensure that each cylinder does not exceed a maximum temperature and that will then control how lean the engine can run before the ECU intervenes. A lean engine will run hotter than a rich engine.
- Fuel pressure sensor – if fuel pressure fluctuates, there will be less force on the back of the injector and a lower pressure will result in less fuel passing through the injector. This is not so important as most fuel systems can easily maintain a constant pressure, but on endurance races other electrical faults could result in a reduced voltage being applied to the fuel pump and, therefore, a reduced pressure. Alternatively, a weak pump or partially blocked filter could also affect pressure.
- MAP sensor – this is sometimes used if a ram air system is used to pressurise the airbox – the MAP sensor will be able to monitor this additional air pressure and the ECU will then increase the injector pulse width to keep the air/fuel ratio correct.

Isolated runner manifolds are the most common set-up used with throttle bodies. This offers excellent fuel/air distribution as each throttle flap and injector will only supply one cylinder of the engine. Fuel atomisation is also highly important because if the air and fuel do not mix properly then there will be an inefficient burn in the cylinders. In low rpm engines, we can fit the injectors close to the engine so they spray fuel on to the back of the valve inside the inlet port (normally via batch injection). Being a low rpm engine, it can have a smaller injector that opens for a longer period of time, which improves atomisation. However, with high rpm engines we have less time to inject the fuel so need to use larger injectors and a smaller pulse width, which reduces atomisation. A solution is to place the injectors further away from the engine, often as far back as the other side of the throttle plate and nearer the air intake trumpets (when using sequential injection). This is beneficial as the fuel droplets will reduce in size as they hit the throttle plate; additional heat from hot engine components and radiated from the engine itself will also aid atomisation. Injecting fuel this way also allows more time for the air and fuel to mix together as it swirls down the inlet tract. If a more efficient burn can be created, the mixture could be made leaner to help increase power.

Twin injector set-ups are also common and are often used by motorcycle manufacturers on their high-revving superbikes. This set-up allows

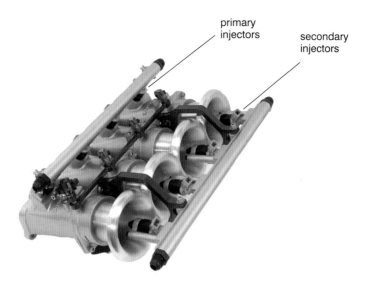

Figure 1.8 Eight-injector set-up

you to use smaller injectors, as mentioned previously, for the benefits of increased atomisation. Using these staged injectors allows for fine-tuning across the whole rev range instead of compromising by using only one size of injector, which will only work well at certain times. Using two small injectors allows for one injector to be situated between the inlet valve and throttle flap – this operates at low to mid rpm. As the revs begin to rise and fuel capacity requirements are increased, the secondary injector, which will normally be situated on the other side of the throttle flap and often in the intake trumpets or airbox, will also begin to inject fuel to maintain good atomisation. This will provide a larger area beneath the power curve across the whole rev range, producing a higher average amount of power. Getting these injectors set correctly will take considerable time on the dyno but will provide you with a responsive and powerful engine across the whole rev range.

Throttle body sizing and selection is important, just like choosing the correct size of carburettor. Relatively small throttle bodies will give you great throttle response and good low-down power but will affect your maximum horsepower output, whereas large throttle bodies will work in the opposite way. Again, there is some overlap to picking the correct throttle bodies, based on the engine's state of tune and how well it breathes. Table 1.2 provides a guide from Jenvey Dynamics to determining throttle body bore based on brake horsepower (bhp) per cylinder, assuming a maximum of 9000 rpm and a distance of 120 mm from throttle plate to valve head.

Table 1.2 Recommended throttle body bore sizes

bhp per cylinder	Throttle body bore size
≤30	30 mm
≤33	32 mm
≤39	35 mm
≤46	38 mm
≤51	40 mm
≤56	42 mm
≤65	45 mm
≤74	48 mm
≤80	50 mm
≤87	52 mm
≤93	54 mm

When selecting throttle bodies, you may use a conventional butterfly throttle flap system or a roller barrel system. Jenvey Dynamics' throttle bodies are one of the most popular systems for the conventional system. Companies producing the roller barrel systems, such as Titan Motorsport and Automotive Engineering, state that this design can increase maximum horsepower as the throttle flap is completely removed at WOT, unlike the conventional system.

You can also specify a tapered throttle body, where the bore is smaller at the point where it reaches the inlet port of the cylinder head. This is ideal for 'direct to head' throttle bodies, which are then easier to match to the ports and allow for a more compact fitment than having an additional inlet manifold to provide the tapering.

Again, a key area to get correct is synchronising the opening and closing of each individual throttle butterfly.

More sophisticated systems can utilise 'drive by wire', which uses a position sensor on the throttle pedal to control a motor on the side of the throttle bodies. This can then give an option to provide a dry and wet weather throttle map to change the characteristics of the throttle application and driveability.

Batch firing of the injectors, when all the injectors spray at the same time with every revolution of the crankshaft, is a very simple system to set up. However, it does mean that the injectors fire when the inlet valve is closed.

Sequential firing is a more modern set-up and allows the injectors to only fire when the inlet valve begins to open.

Fuel-supply system

This contains the fuel injectors, a pressure regulator, fuel and fuel filter. There is also the fuel tank itself and, dependent upon the application, a fuel swirl pot to avoid fuel starvation if the tank is not sufficiently baffled.

Fuel injectors are all rated at a specific flow rate at a given fuel pressure (e.g. 450 cc/min at 2.7 bar). This value does not hold much relevance though as the amount of fuel that ends up going into the inlet system is determined not only by injector

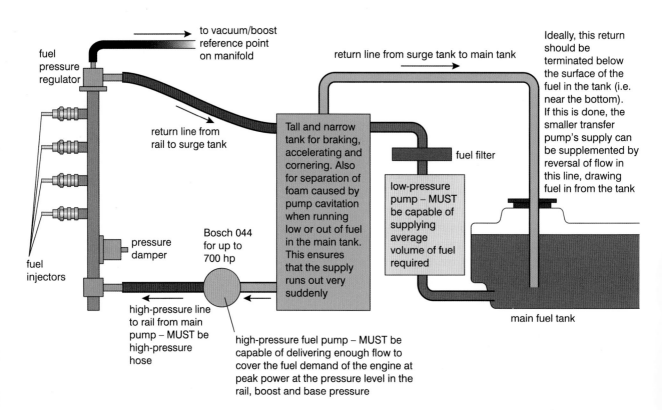

Figure 1.9 Fuel-supply system

size, but also by the fuel pressure behind it and its pulse width.

An injector's duty cycle is the maximum percentage of time the injector is spraying fuel to meet the engine's maximum fuel-flow needs at maximum power. If an injector is active all the time, it would eventually fail due to the solenoid inside the injector body overheating. A duty cycle range of 60–80 per cent is often common to ensure that the pulse width is long enough to correctly atomise the fuel but not long enough to damage the injector. A reduced duty cycle of course then reduces the flow rate of the injector (i.e. 450 × 60% = 270 cc/min).

To calculate the approximate fuel flow needed per cylinder for a petrol engine, we can assume the following calculation:

$$\text{Fuel flow (cc/min)} = \frac{HP \times K}{C}$$

Where:

HP = Maximum horsepower

K = 4.6 for naturally aspirated engines and 5.6 for forced induction engines

C = Number of cylinders

If we assume a naturally aspirated four-cylinder engine at 250 bhp, we end up with:

$$\frac{250 \times 4.6}{4} = 287.5 \text{ cc/min}$$

From there, we can then look at the injector size we would need when taking into consideration our duty cycle.

$$\text{Injector static flow (cc/min)} = \frac{TF \times 100}{N \times M}$$

Where:

TF = Theoretical flow

N = Number of injectors per cylinder

M = Injector duty cycle

If we assume we are running sequential injection with one injector per cylinder with a duty cycle of 80%:

$$\frac{287.5 \times 100}{1 \times 80} = 359.4 \text{ cc/min (to 1 d.p.)}$$

This then specifies what size injector you will need, or you have the option of purchasing a smaller injector and running it at a higher fuel pressure. So if we had a suitable injector that was rated at 300 cc/min when rated at 2 bar by the manufacturer, we could look to increase fuel pressure to 3 bar and then use a smaller injector in line with the following calculation:

$$\text{Revised static flow} = SF \times \sqrt{\frac{RP}{OP}}$$

Where:

SF = Static flow

RP = Revised fuel system pressure

OP = Manufacturer's stated pressure

So:

$$300 \times \sqrt{\frac{3}{2}} = 367.4 \text{ cc/min (to 1 d.p.)}$$

This should give you an idea of the approximate size of injector you will need. However, you can always have them flow tested at different pressures to match different size injectors and pressures to a specific engine. This can also ensure that the injectors you use all have a flow rate within a few per cent of each other.

Other areas to consider:

- Fuel pressure regulator – this item allows the system to maintain constant fuel pressure and is normally found on the end of the fuel rail or on the return line. Some fuel pumps, such as the one found on a Suzuki GSX-R1000, have a fuel pressure regulator built in to the fuel pump. This is compact, but is a long way from where the pressure needs to be measured from (fuel rail). There are three different types of fuel pressure regulator:
  - Fixed pressure – this is rated at set pressure and, once fitted, cannot be adjusted.
  - Adjustable pressure – this will have some kind of adjustment to allow you

to manually control the fuel pressure within a set of parameters.

- Vacuum controlled – this uses a take-off pipe from the inlet manifold to allow the fuel pressure to be increased as the manifold pressure increases.
- Fuel pump – it is important to specify a fuel pump that will over supply the engine from your previous calculations as you must assume that it will spend some time at WOT at near maximum revs. This will then allow you to be sure that enough fuel is being supplied to be able to maintain a constant fuel pressure. Some endurance cars will use twin pumps to allow them to switch over if one fuel pump fails.
- Fuel filter – a high-flow fuel filter with a replaceable element is a good option, however, always ensure that the directional arrow is in the direction of fuel flow. There are a range of filters you can fit; the glass type enables you to easily inspect the fuel filter without dismantling any of the fuel lines.
- Swirl pot and fuel lines – with basic fuel tank designs and cars capable of producing high acceleration forces (gs) in cornering, accelerating or braking, you may need a swirl pot to control fuel surge and starvation. As the fuel level becomes low, fuel can move away from the fuel pickup and cause starvation – a way around this is to fit a swirl pot. You will need a fuel pump to supply the swirl pot, and then the pressure pump to pull fuel from the swirl pot up to the engine. Braided lines are the best choice to avoid any catastrophes from chaffing or vibration. Avoiding too many angled pipes and routes will also help to avoid flow restriction and working the fuel so that it begins to heat up.
- Fuel tank – there are many designs for tanks: bag tanks are the most expensive but safest. Baffles and foam are commonly fitted to tanks to avoid fuel surge, while a one-way breather must be fitted to release air pressure in the tank and to avoid any spillage if a rollover occurs.
- Fuel cooler – not a commonly used addition, but keeping the fuel cool helps to keep it more dense.

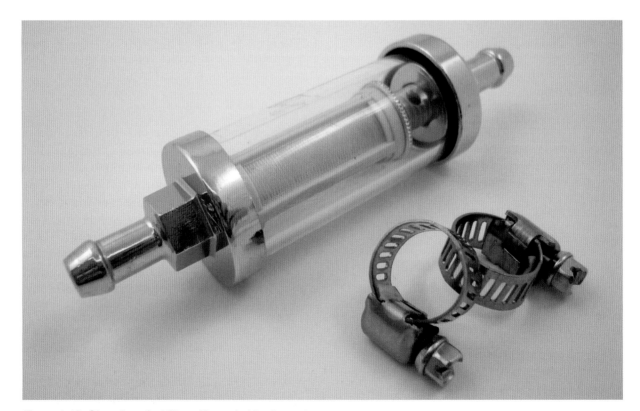

Figure 1.10 Glass-type fuel filter with washable element

1.4 Ignition system

The ignition system is an important part of the engine as it controls what is done with all of the fuel and air that is crammed into the cylinder. It needs to burn the whole mixture at the right time in order to provide power. The burning speed must be controlled because if it burns too quickly after TDC, the engine will not run smoothly and will eventually fail after prolonged use, due to excessive loadings on the components when pressure on top of the piston is very high. If the burn is too slow, it will not produce the power required, as the force on the piston will not be strong enough. Once the spark has ignited the mixture, factors such as mixture quality, movement/turbulence and the design of the combustion area all affect the way the mixture will burn.

Assuming you have an understanding of ignition systems, you will know that there are three main systems in use (oldest first):

- Mechanical distributor
- Electronic distributor
- Electronic ignition (distributorless)

Ignition advance and retard are the most commonly discussed areas of the ignition system. The idea behind this principle is to allow the mixture the correct amount of time to burn properly. As an engine begins to rev faster, we must advance the ignition so that the peak cylinder pressure forces the piston down the cylinder at around 12° after TDC. As the engine

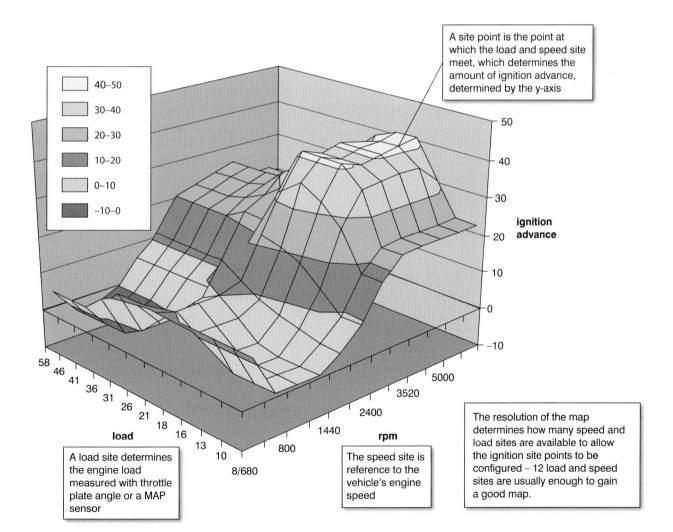

Figure 1.11 Three-dimensional ignition map

revs change, so must the ignition timing in order to keep the engine producing maximum power across the whole rev range. For example, at high rpm and load, we may have an ignition advance of around 30° before TDC (BTDC), while at low rpm and load, this will be retarded back to around 10° BTDC, as the engine is moving slower.

All engines are different as burn rates vary greatly based on air/fuel ratio, mixture density, fuel type and engine design. All of these can be affected by the intake/exhaust design, cam type, compression ratio and so on.

Engines fitted with distributors have a limitation as not all rev and load ranges can be perfectly tuned, so some areas will have to be a compromise when initially set up.

The electronic ignition system is the best for all scenarios as it allows a three-dimensional map to be built so that ignition advance can be adjusted based on varying rpm and engine load. A programmable ECU allows this to happen – the map will build up using an engine dyno with a laptop plugged into the ECU. The tests are carried out at multiple rpm and engine load parameters to gain the best map for maximum power throughout the rev range. With advanced ECUs and a set of exhaust gas temperature sensors, you can also have slightly different maps for each cylinder, as discrepancies in machining, inlet/exhaust systems, cooling jacket locations, and so on, may mean that the standard map will not be optimum for every cylinder in the engine.

1.5 Cylinder head

The cylinder head is hugely important in the engine system. It has a significant effect on how much air and fuel enter the cylinder and in what state they are in when they arrive. The head needs to have a high airflow rate (which can be measured using a head flow bench) and, in turn, must also assist in mixing the air/fuel mixture allowing it to burn efficiently. Creating maximum airflow is important but does not always result in a higher power; as stated earlier, the mixture and efficiency of the burn must also be considered.

The cylinder head and its components in general can be split into the following categories:

- Inlet valve and port
- Exhaust valve and port
- Camshaft and valve train
- Combustion chamber and cylinder head ancillaries

1.5.1 Inlet valve and port

Modifying inlet valve size is a common and effective way to improve engine performance. An increase in valve diameter may only be small but, when considering airflow into the cylinder, the lift of the valve multiplied by the valve area shows how much more additional air will be able to enter the engine. For example, with 10 mm of valve lift, the difference in volume of space between a 32 mm diameter valve and 33 mm valve is 510.7 mm^3. Engines with more than two valves per cylinder tend to have a good advantage in the fact that they can fill the combustion chamber roof, maximising the volume or mixture that is able to enter the cylinder. However, bigger is not necessarily better, as the larger you go with valve sizes, the more you tend to lose the low-end performance of the engine due to poor control of gas speed and mixture. Titanium valves are a common upgrade to lighten the valve train (by around 30 per cent) and reduce reciprocating mass (mass that rotates or moves). Undercut valve stems also allow for increased flow, as the part of the stem that is exposed in the port is reduced in diameter and causes less restriction – this is something that must be tested first as valve fatigue could easily see the valve head drop into the cylinder and cause damage.

The inlet port itself is important – circular inlet ports tend to create less turbulence and drag. Most engine builders keep port diameter at a maximum of around 85 per cent of the valve diameter in order to keep a good balance between airflow and velocity across an acceptable rev range. The port must add swirl to the incoming mixture in order to allow it to mix properly and provide efficient combustion.

Swirl can also help to distribute any remaining unscavenged exhaust gas instead of leaving it as a pocket in the cylinder. A curved port will promote swirl and allow all the fuel molecules to be broken down much more, which will then mix better with the air. As the mixture moves towards the back of the valve, it is important to ensure that any radius changes are gradual so that the mixture can stay attached to the port walls. Entry into the combustion chamber must stay smooth and progressive while maintaining swirl – to do this material can be added or removed. The inlet port should be matched to the manifold in order to provide smooth flow.

The valve seat can also be enhanced to increase flow. Most standard valves and valve seats have a single 45° angled cut to allow the faces to mate together. However, this individual angle is not good for airflow. Engine builders, with access to specific machining tools, can cut three or more angles into the seat and valve and blend these together to increase flow. The progressive cut normally utilises a 30–45–60° cut, which helps to steer the mixture in the correct direction. However, with the use of extremely high accuracy machines, some builders are now using five or more angles. Valve seat sealing is very important in order to maintain good compression and power strokes.

Figure 1.12 Worked cylinder head

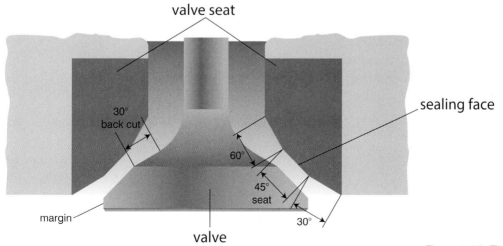

Figure 1.13 Three-angle valve seat

1.5.2 Exhaust valve and port

The exhaust valve is often around 70–85 per cent of the size of the inlet valve on a race engine for a few reasons. Firstly, the inlet valve takes priority in order to get as much air into the engine as possible. Secondly, a correctly designed exhaust system should be able to scavenge the burnt gases out of the engine when the exhaust valve begins to open. As the exhaust valve is exposed to high temperatures, undercutting the valve stem is not popular as it may risk the reliability of this component. Sodium filled valves allow greater heat transfer out of the ports and to the cooling jackets. The hollow valves contain sodium, which melts during engine operation. Valve action causes the sodium to circulate, removing heat from the valve head. The heat travels up the valve stem and is transferred to the cylinder head.

The exhaust ports tend to be around 90–100 per cent of the size of the valve in order to increase gas flow out of the cylinders. Flow of the exhaust gases is very efficient with a D-shaped port (the flat section being the floor of the port). The exit of the port should be smaller than the exhaust header in order to maintain mid-range power; this reduces the gas backflow once it has exited the port.

The exhaust valve seat is not only a gas tight seal, but also a key area for cooling the exhaust valve as the direct contact with the cylinder head allows the coolant to pull heat away from this hotspot.

1.5.3 Camshaft and valve train

The camshaft is another key area of the cylinder head assembly (if we assume an overhead cam set-up). It controls the operation of the inlet and exhaust valves in terms of valve lift and duration. The 'lift' is how far the valve is opened at its maximum point, and the 'duration' is how long the valve is actually open. The relevant valves will open just before a stroke starts and close just after it finishes. It works this way so the fast-moving gases that build up momentum have every chance of entering and filling the cylinder.

The inlet valve begins to open before TDC so that the piston is already off its seat as the induction stroke begins. It will also stay open as it passes bottom dead centre (BDC) as the high gas velocity allows the mixture to continue to rush into the cylinders. The exhaust valve opens before the end of the power stroke (as most of the energy has been expelled halfway down the power stroke), reducing pressure on the piston on its way back up. At the top of the exhaust stroke, the exhaust valve stays open for a short period after TDC, to allow the exhaust to scavenge any remaining gases as well as to create a vacuum that helps force the inlet charge to be sucked into the cylinder. This period is known as valve overlap – when both valves are open together during crossover of strokes.

This works well at high engine rpm but not very well at low rpm when the gases travel at slower speeds. Sadly, a camshaft is a compromise that will not work perfectly from low to high rpm, which is why some engines are designed and produced without camshafts to get over this problem; although cam-less engines also have their own pitfalls.

A high-performance cam has very subtle differences to a standard road cam; the differences are in the shape of the cam lobes. Cam lobes have four parts:

- Base circle – this keeps the cam on its seat and allows it to be rested and cooled; as a result, it must be at a constant radius from the centre of the cam.
- Ramp – this controls the initial opening and closing of the valve.
- Flank – this controls the opening and closing of the valve.
- Nose – this is the tip of the camshaft lobe and determines the maximum lift characteristics of the valve.

Table 1.3 Full race camshaft specification

Application	Power band (rpm)	Duration		Valve lift		Timing (degrees)	Full lift	
		Inlet	Exhaust	Inlet	Exhaust		Inlet ATDC*	Exhaust BTDC†
Designed as a full race/sprint cam where maximum power and top end torque are required	4000–8500	300°	300°	0.462 inches 11.73 mm	0.435 inches 11.05 mm	48–72 84–56	102°	104°

Table 1.4 Fast road camshaft specification

Application	Power band (rpm)	Duration		Valve lift		Timing (degrees)	Full lift	
		Inlet	Exhaust	Inlet	Exhaust		Inlet ATDC*	Exhaust BTDC†
This fast road cam gives a good torque and power increase in road cars	1500–6700	264°	264°	0.400 inches 10.16 mm	0.400 inches 10.16 mm	22–62 62–22	110°	110°

*ATDC = After top dead centre
†BTDC = Before top dead centre

A pair of camshaft specifications for a 2.0 litre 16V engine can be seen in Tables 1.3 and 1.4. The difference in performance between the two can be clearly seen: the power bands are completely different in terms of the rev range that they operate best within, and the race cam has a longer duration and more lift on both the inlet and exhaust cam. The timing for the race cam is 48–72/84–56, which means the inlet valve opens 48° BTDC and closes 72° after BDC, while the exhaust valve opens 84° before BDC and closes 56° after TDC. The angle at which full lift is reached can also be seen in the final column.

Dependent upon the amount of lift being used, you may have to use pistons with valve cut-outs in order to stop the valve and piston from

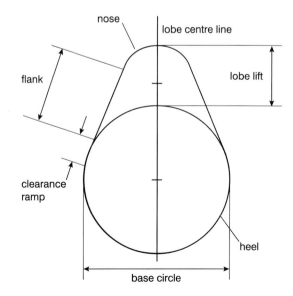

Figure 1.14 Cam lobe diagram

clashing. The specification of the camshaft, based on its duration, lift and overlap has a huge bearing on the engine's torque and power curve along with its driveability.

Camshaft timing is easily achieved with a set of vernier pulleys or offset dowels, as these allow the cam valve opening and closing to be advanced or retarded. This effectively allows you to tune the cam further. By advancing the cam timing, the valves will open and close earlier, while retarding the cam timing means that the valves will open and close later, allowing for further tuning of the power band and control of valve overlap, cylinder pressure and how the inlet and exhaust valves control the gas flow. Advancing the cam will often give better bottom-end and mid-range power, while retarding it may give slightly better top-end power at the sacrifice of the bottom-end and mid-range.

Selecting the correct cam is vital but can be difficult. You must ensure that the decision is based on your engine package. Picking the full race cam outlined in Table 1.3 would be pointless if you only have a stock bottom end and the maximum safe rpm of the engine is 7500 rpm. Equally, picking the fast road cam would effectively strangle the engine if you have a worked set of ports, together with large throttle bodies and a large bore exhaust.

Cam manufacturers can measure their cam duration and lift in different ways so it is important to understand this. For example, some will measure direct movement of the cam bucket, while others will measure from actual valve movement and take into account valve train flex and so on. For this reason, two cams with the same specifications from different manufacturers may have completely different characteristics.

Aggressive cams and high rpm can also be subject to valve bounce and toss. Valve bounce is when the valve violently slams shut against the seat (normally due to high duration angles) and, instead of staying there, it rebounds back off the seat as there are no damping forces – this can then oscillate for tens of degrees before coming to rest. Valve toss is when the valve is flicked off the cam at around full lift as the cam loses contact with the follower due to the mass of the valve assembly overcoming the valve spring. This loss of valve control is detrimental to the valve train due to the shock loading being forced through it, and control of engine performance becomes difficult to predict. Valve toss is not as detrimental, as it can also often give a small performance increase, but valve bounce can ruin compression and intake strokes. Increased valve spring rate and a lighter valve train are a good way to overcome this if the cam aggression and engine rpm cannot be reduced.

High rpm engines (≥ 7500 rpm) will normally have solid lifters fitted, as hydraulic lifters cannot keep up with the constant need to build back up to full size with oil pressure, as the time between cycles is reduced as the engine revs

Figure 1.15 Vernier cam pulleys

harder. The lifters provide the link between the cam and the valve itself. They have a huge amount of force applied to them and so their spherical radius on their top surface allows the lifter to rotate in order to reduce wear. Valve clearance is controlled by using shims between the end of the valve stem and the underside of the lifter.

Pushrod engines utilise a rocker system, which normally has an adjustment screw to alter valve clearances and the ability to adjust the rocker ratios thereby altering valve motion characteristics without changing the camshaft.

Valve springs are vitally important to engines that rev highly and are fitted with aggressive cams. They ensure the valve follows the cam's path and then stays shut against the valve seat. Using the right spring set is important in order to avoid coil bind, power loss and valve train wear. Remember that when you increase the lift of the cam, the spring will have to compress further and the valve spring seat may need to be modified to stop them from binding. On any competition spring being installed, it is also vital that all valve springs are fitted with the correct seat pressures.

When fitting camshafts, it is a good idea to use a product such as Graphogen Assembly Compound to lubricate the bearings, lobes and lifters to avoid any premature wear.

1.5.4 Combustion chamber and cylinder head ancillaries

Volumetric efficiency refers to how well the cylinder is filled with combustible mixture and is an important design factor of an engine. A normally aspirated engine has a volumetric efficiency of approximately 80 per cent, meaning that for a normal induction stroke the cylinder is filled to approximately 80 per cent of capacity. So a single cylinder engine with a capacity of 100 cm^3 would be filled with 80 cm^3 of air and fuel in a normal stroke.

It is known that the amount of airflow that travels through an engine is proportional to the power that it will produce (as well as the other factors already discussed), so an increase in airflow or volumetric efficiency will lead to an increase in power. It is possible to increase volumetric efficiency of an engine in the two ways outlined below.

Normally aspirated engines

Increasing the airflow through modifications to the camshaft, induction and exhaust system will increase volumetric efficiency. By modifying the camshaft the duration for which the valves stay open can be increased, which enables the engine to draw in more air than would normally be possible. Modifications to the induction system can increase flow of air into the cylinder or even force more air into the cylinder when pulse tuning is used (by adjusting the inlet runner length). The exhaust system can also be tuned to scavenge the burnt gases out of the cylinder to create more of a vacuum in the cylinder during valve overlap. Using these methods together can produce a volumetric efficiency up to around 100–105 per cent, meaning that the cylinder mentioned above could well hold more than 100 cm^3.

Forced induction engines

These are engines that use a turbo or supercharger to force air into them. This increase in air in the cylinder causes an increase in pressure, allowing volumetric efficiency to reach 130 per cent or more.

Volumetric efficiency is calculated by dividing the 'brake airflow' (the actual airflow through the engine) by the indicated airflow (the theoretical airflow through the engine). The actual airflow is measured, usually using an airflow meter, and the theoretical airflow is calculated, using the engine speed and capacity.

$$\text{Volumetric efficiency} = \frac{\text{Brake airflow}}{\text{Indicated airflow}} \times 100$$

$$\text{Brake airflow (l/s)} = \frac{\text{Mass airflow per cylinder (g/s)}}{\text{Relative density of air}}$$

Indicated airflow (l/s) = Number of working strokes per second × Displacement of cylinder

With a four-cylinder 2.0 litre engine that produces maximum torque at 2500 rpm with a mass airflow of 47.25 g/s, we can calculate volumetric efficiency. Relative air density is assumed to be 1.225 g/l.

$$\text{Brake airflow} = \frac{(47.25 \div 4)}{1.225} = 9.643 \text{ l/s}$$

$$\text{Indicated airflow} = \frac{2500}{2 \times 60} \times \frac{2}{4} = 10.42 \text{ l/s}$$

$$\text{Volumetric efficiency} = \frac{9.64}{10.42} \times 100 = 92.5\%$$

Compression ratio is the relationship between the volume above the piston at BDC compared with the volume above the piston at TDC. Road engines normally have a ratio around 10:1, while race engines can often reach high levels of up to 13:1. The higher ratio increases the pressure of the compression stroke across the whole rev range. Increasing the compression ratio increases the thermal efficiency of the engine. The thermal efficiency is a measure of how the engine converts the heat produced into mechanical power. The compression squashes the air and fuel together to further increase mixture quality. However, with an increase in compression comes an increase in temperature. Standard pump fuel (garage forecourt fuel) has an upper temperature limit, before detonation occurs, that would be unsuitable. Fuel with a higher octane rating (≥100) must be used, as a lower octane fuel is more likely to ignite spontaneously in the increased temperatures.

To measure the compression ratio requires great accuracy, using the following equation:

$$CR = \frac{CV + CCV}{CCV}$$

Where:

CR = Compression ratio

CV = Cylinder volume/swept volume (capacity of the cylinder)

CCV = Combustion chamber volume/clearance volume

The volume of the combustion chamber includes any space that is between the piston at TDC and the face of the combustion chamber, and the volume that the head gasket would take up. You will need to determine what pistons are being used, as some are domed, some are flat topped and others are dished. The flat top is easier to calculate but you must subtract the volume for a domed piston, and add the volume of a dished piston.

A burette filled with fluid is a common way of measuring the cylinder and combustion chamber volumes to help calculate the compression ratio. With the valves closed, you can put a glass plate with a hole in the middle on top of the combustion chamber and seal it with grease. Place the burette in the hole and allow the fluid to fill up the chamber, noting how much fluid has entered the chamber.

For example, a 2000 cc engine has a cylinder volume of 500 cc and the space above the piston at TDC (including gasket and combustion chamber) is 50 cc (assuming it has a flat top piston) – this would give us the following equation:

$$\frac{500 + 50}{50} = 11$$

Compression ratio = 11:1

Head gasket clamping force is very important, due to the high demands of stress that the engine is placed under, along with ensuring that both block and head surfaces are flat. Engine detonation (spontaneous combustion) can also damage the gasket. Head studs clamp more effectively than bolts and cause less distortion when tightening. Some builders also drill out the head and block to increase the size of the studs or bolts, which allows for a greater clamping force. The tightening sequence of the head must be strictly followed and it is often best to tighten four or five tightening sequences so that the head gets tightened evenly to distribute the load.

1.6 Exhaust system

The exhaust system is a key part of the engine package; when correctly designed, it can be used to scavenge the gases out of the cylinder and, when valve overlap occurs, it can also help draw the mixture into the cylinder, improving volumetric efficiency.

Assuming a four-cylinder engine, the options for an exhaust system are quite varied. You must consider the bore of the exhaust and whether this stays constant, header lengths, type and amount of collectors (4–1 or 4–2–1 system), tailpipe length and diameter, fitment to cylinder head and silencer.

Gases leaving the engine build up momentum and the inertia that is created aids the exit of gases. The flow continues after the valve has shut, which then creates a depression at the beginning of the exhaust. The next time the valve opens, the gases rush from high pressure to low pressure (cylinder to exhaust). Using an individual header (or primary) per cylinder allows this to occur.

The four primaries joining together at some stage enhances the engine's performance by increasing power and stretching the power band. From the four primaries, a collector decreases the number of pipes. This can be to two secondary headers or it can go straight to the tailpipe.

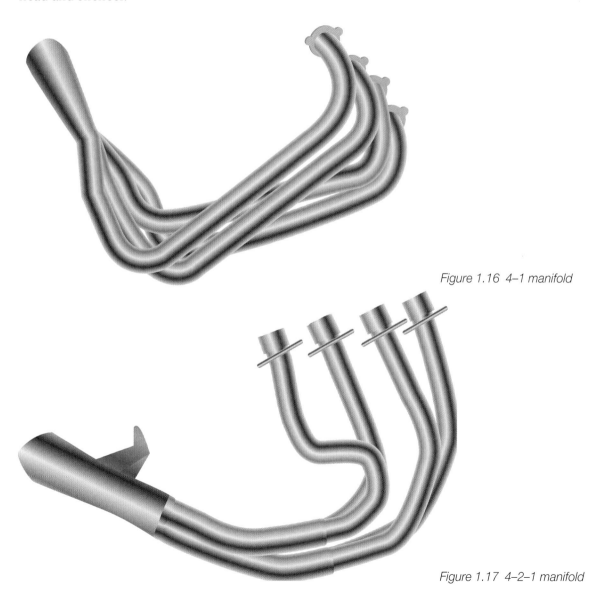

Figure 1.16 4–1 manifold

Figure 1.17 4–2–1 manifold

A 4–1 system can give a better top end but normally weighs more due to the need for four long primaries, whereas the 4–2–1 system is more compact, weighs less and will give a better mid range.

Pulse tuning is used so that the pressure waves are used effectively to improve cylinder scavenging and assist with cylinder filling. As the pressure waves exit the exhaust pipe, they cause a negative pressure wave behind them, which works its way back towards the engine. Using the correct length and diameter exhaust system will allow that negative pressure to arrive at the back of the exhaust valve when valve overlap occurs, which means the exhaust gases will be scavenged and the fresh mixture will be sucked into the cylinder.

If we can get the exhaust to work harmoniously with our cam and inlet, it is possible to see large power gains. Without this synchronisation, during valve overlap, you could end up with no scavenging and, as a result, poor gas flow and mixture quality.

The basic principle of sizing an exhaust is that the bore of the primary will determine where peak torque occurs, while the length of the primary determines where the power curve sits in relation to peak torque. Longer pipes increase low-down and mid-range power, while shorter primaries increase top-end power at the expense of low-down and mid-range power.

To calculate primary pipe length, we can use the following equation:

$$P = \frac{850 \times ED}{rpm} - 3$$

Where:

P = Primary pipe length

ED = Number of degrees the exhaust valve opens before BDC

rpm = Engine speed that the exhaust is being tuned to

For race engines, an rpm near maximum horsepower or a little less is often used. Using a four-cylinder, 2.0l engine, a tuned rpm of 7500 rpm and an ED of 74° we can calculate primary length to be:

$$\frac{850 \times (74° + 180°)}{7500} - 3 = 25.8 \text{ inches}$$

This is the set length for each primary, which is why you see that each individual primary will have different bends and routes in order for them to equal the same length when they enter the collector and to have an easy route for the gases to flow without restriction.

To work out the internal diameter (ID) of the exhaust we can use the following equation:

$$ID = \sqrt{\frac{cc}{(P + 3) \times 25}} \times 2.1$$

Where:

cc = Cylinder volume

P = Primary pipe length

Taking an example of 500 cc, this would look like this:

$$\sqrt{\frac{500}{(25.8 + 3) \times 25}} \times 2.1 = 1.75 \text{ inches}$$

When using a 4–2–1 exhaust system, the calculations are the same except the ID of the second primaries is worked out as follows:

$$ID = \sqrt{(ID^2 \times 2)} \times 0.93$$

This then can be worked out to be:

$$\sqrt{(1.75^2 \times 2)} \times 0.93 = 2.4 \text{ inches}$$

It is worth remembering that we are, of course, restricted to the ID pipe we can buy from the suppliers, and that the length of the second primaries includes the length of the first collector.

With a 4–2–1 system, the balance of the engine's power can be altered with differing lengths of primaries (longer gives more top-end power) and secondary primaries (longer gives more mid-range power).

The set of calculations on the previous page should give you a good base to start from but every engine is different in terms of head work, flow and mixture characteristics. However, once a dyno run has been carried out and the torque and power curve has been analysed, you can go about adjusting diameters and lengths to suit your application.

Different lengths of primaries across the cylinders are also used, based on the theory that instead of the engine producing one high amount of peak torque we can have a variety of primary lengths, which may end up producing less peak torque but more torque over a broader range, as each cylinder will produce peak torque at a slightly different rpm. Again, dyno testing will show you how it affects each engine, but you should start by varying each primary by around 1 inch from the current length and build from there.

Stepped primaries are another concept in which there are one or two increases in the primary diameter of about 1/4 inch as it reaches the collector. What this allows, is the engine power band to be further stretched, but again this will demand more dyno time. Time on the dyno would allow you to work out where the steps should be placed in order to remove any holes in the power band while keeping maximum power where you want it – however, this costs money!

One or more collectors come after the primaries and this is the part where those primaries become secondary primaries or the tailpipe. The collector has the ability to tune below the peak torque and there are a multitude of designs to choose from. The collector joins the primaries together, allowing for any unused negative pressure wave in a primary to be used by another primary. There are four main types of collector: baffle, merge, venturi merge and split interference.

The tailpipe and silencer are the final part of the exhaust, which allows the gases to expel to the atmosphere. The tailpipe will generally be about the same length as the primary lengths, so a 2.0 litre engine with 250 bhp will have a tailpipe ID of approximately 3 inches in order to allow the gases to exit freely without any restriction but to also keep the gas speed up to a reasonable level.

Figure 1.18 Repackable silencer

Noise restrictions generally dictate silencer size and, while they cause a power loss, they are essential to pass the noise test and get on to the circuit. Every engine design is different and in most cases a standard size can is fitted, but with some more dyno time, the same noise absorption can be found across a range of silencer sizes and designs, but all will have different losses. Finding the right one to match your car will reduce the loss of power. One solution is wire wool around the perforated tube and then basalt in the silencer. This lasts longer and is more consistent at noise absorption than E-glass. E-glass also turns into a crystal-like structure once it gets overheated. Having a repackable silencer allows you to remove the end cap (use stainless steel riv-nuts and bolts) and inspect or repack the silencer. There is a relatively new product that is also becoming popular known as Acousta-Fil, which has got great reviews.

Stainless steel is the most commonly used material for exhausts due to its strength and resistance to corrosion. Titanium is a more expensive option but can give a weight saving of around 40 per cent. Inconel is stronger and more durable than titanium, so a thinner-walled tube can be used, saving more weight, and it also retains heat very well, giving it a greater advantage when used in areas of tight packaging. Professional exhaust fabricators will purge the exhaust with gas while welding the tubes together in order to stop the weld from protruding into the inside of the exhaust – this ensures that the flow does not create any unnecessary turbulence or restriction between pipe joins and joins to the mounting flange.

When fitting the exhaust to the car, ensuring a secure mount is key – an anti-vibration mount along the exhaust may be used to reduce the risk of fracturing the exhaust.

1.7 The bottom end

The bottom end can be considered to be everything below the cylinder head, including the block itself, pistons, conrods, crankshaft, and flywheel.

This part of the engine usually remains untouched until we start to move towards the semi-race and full-race types.

The block houses the cylinders, oil and water galleries so block preparation is important before building the engine to ensure that it is true and free of any cracks, signs of damage or wear, and that threads are in good condition and galleries are free of any blockage. Bore alignment to the crankshaft is important, along with ensuring that they are honed by engine builders (further improving cylinder trueness and ring sealing properties). Cylinder bores can also be plated, reducing surface friction between the cylinder walls and the piston rings, using a coating such as Nikasil. With a reduction in friction not only comes a small power gain, but also reduced wear of the components involved. Nowadays, all cylinder blocks for racing applications are honed with the deck plates fitted for perfect roundness when cylinder heads are refitted and main bearing caps are torqued up at the same time, so when the engine is assembled it is the same as if it were running.

When boring the cylinder to a larger size to fit larger pistons and, therefore, increase engine size, it is important to consider cylinder wall thickness, as an engine bored out too much will result in the walls warping due to the heat and pressure from the cylinder, which could result in poor cylinder performance and potential issues with engine reliability.

Boring a cylinder will provide a boost in power so long as the top of the engine can manage the additional mixture flow required to fill the larger cylinder capacity. The compression ratio must also be checked to ensure that it does not become too high.

The pistons will, of course, need to be replaced if the cylinders become larger. Forged pistons will need to replace the standard cast versions in order to cope with high cylinder pressures. Forging produces a higher density piston and one that has a higher tensile strength and a better thermal conductivity. Billet pistons are also available, which may be slightly lighter and

Figure 1.19 Slipper piston

stronger but, at the same time, the cost increase may not be worth the small gains over a forged piston.

Racing pistons can be full skirt, slipper or a mix between the two. The slipper piston is lighter, whereas the full skirt piston will normally have better ring sealing, oil control and durability. The top of the piston can come with a variety of different designs to suit combustion chamber shape and size, to alter compression ratio and to provide clearance for the valves.

The piston ring(s) form the seal between the piston and the cylinder wall to keep the gases above the piston and to also act as a bearing that must be durable. Usually three piston rings will be used; the top one being the compression ring, the middle ring to assist the top ring and to also help the bottom ring, with the bottom ring being the oil scraper ring. The top two rings also have to help dissipate the heat from the piston to the cylinder walls. You may also find that some high-end pistons only contain two rings, one compression and one scraper.

Conrods are the next part of the bottom end, connected to the piston with a gudgeon pin and to the crankshaft via the big end bearing. The conrod has to deal with a lot of force, from varying cylinder pressures to changing tensile loads from going through the four strokes continually at high speeds. These are often the most common components to fail in an engine due to the constant switching of loads.

Most competition rods are forged from carbon steel or a powdered form of metal but high-end rods can be made from titanium. They will consist of two pieces: the rod itself and the big end cap, which will be a machined or fractured fit to the rod. Rod strength is determined by a number of factors: material purity, machining accuracy, heat and surface treatments, and the shape and weight of the conrod.

Conrods can be found in two basic shapes: an 'I' beam or 'H' beam. There is a variance of opinion as to which one is better, based on strength and durability, but the other factors previously mentioned are likely to have a greater impact on performance.

Figure 1.20 H-beam and I-beam conrods

The next part to consider is the crankshaft, which is normally the last to be upgraded but is the most important when rev limits are raised when a full race engine is built. Once the rev limit is raised, it may be wise to replace the cast iron crankshaft for a forged steel option as the higher density created by forging gives higher strength and better fatigue resistance.

Billet crankshafts are available, but at a higher cost.

Stroker cranks can be used to increase engine capacity by increasing the distance between the main journal and big end journal. This normally increases torque but can reduce the safe rpm limit slightly. To allow this to work, the length of the conrod must be reduced.

To calculate engine capacity, you need to know the bore and stroke of the engine. Taking an engine with a bore and stroke of 86 mm, we can use the following equation:

$$\text{Engine capacity} = \frac{\text{Radius of bore}^2 \times \pi \times \text{Stroke} \times \text{No. of cylinders}}{1000}$$

So this would be:

$$\text{Engine capacity} = \frac{43^2 \times \pi \times 86 \times 4}{1000} = 1998.2 \, \text{cc}$$

The balance of the bottom end must be focused on so that vibrations do not cause damage to the engine. It is important to match the weight across each cylinder of the rods (big ends and small ends separately) and pistons, followed by balancing the crankshaft to suit.

1.8 Lubrication and cooling

1.8.1 Lubrication

Race engines often require a more complex oil system than the one fitted as standard due to the large increase in accelerations through cornering, forward acceleration and braking (known as g-force). With the increase in the g-force that the engine is exposed to, oil starvation is a common problem and surge/starvation can cause extreme engine damage.

The following systems can be adopted for race applications (performance and cost increasing down the list):

- Wet sump (standard)
- Baffled wet sump
- Swinging sump
- Oil accumulator
- Dry sump

Wet sump

A wet sump is driven by a mechanical oil pump, which sucks up oil from a large (often quite tall) sump pan at the bottom of the engine and pumps the oil around the engine to the necessary parts. A pressure relief valve is used to limit the oil pressure as engine speeds rise so that oil pressure can be lost back to the sump.

A wet sump is a standard fitment. It does not protect against oil surge and can isolate the pickup. As it is tall, it affects the car's centre of gravity (CG) and also does not allow a large quantity of oil to be carried. It is a fairly compact and simple set-up, at a low cost.

Baffled wet sump

A baffled wet sump is an additional machined plate that sits in the original wet sump, helping to reduce oil slosh to keep it nearer the pickup pipe. It is a cheaper alternative to the other systems mentioned. Some people slightly overfill the engine by up to 0.5 litres in a race/

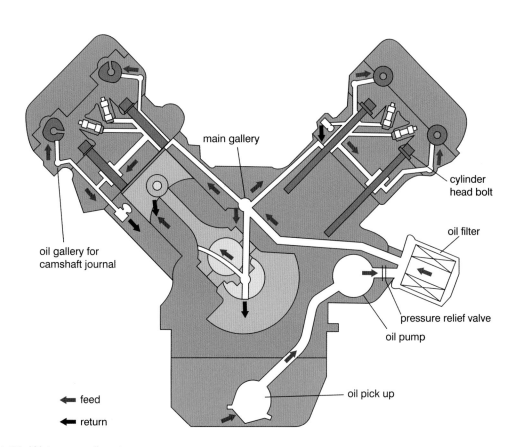

Figure 1.21 Wet sump oil system

Figure 1.22 A sump baffle

track application to cope with the g-force. This system is not advised in a full race application – more for track day use.

Swinging sump

The swinging or rotating pickup sump is a viable alternative to a full dry sump system. It uses the centrifugal force generated when cornering to cause a freely rotating weighted pickup arm to follow the oil in the opposite direction to the corner being negotiated. It can be used alongside a baffle plate.

Oil accumulator

An oil accumulator is a cylindrical aluminium storage container that acts as a reservoir of pressurised oil, to be released when there is a drop in the oil pressure. It is connected to the pressure side of an engine's oiling system and is charged by the oil pump. Its simple, efficient design revolves around a hydraulic piston separating the air pre-charge side and the oil reservoir side. On the oil side, it has an outlet that goes into the engine's oiling system, controlled by a valve. On the air side it is equipped with a

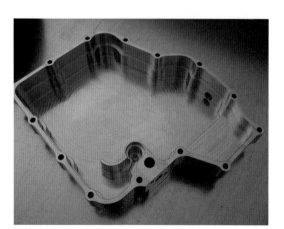

Figure 1.23 Swinging sump showing the pan on the left and the rotatating pickup arm on the right

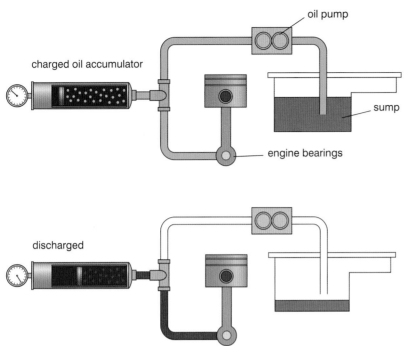

Figure 1.24 Oil accumulator system

pressure gauge and a Schrader air valve, which allows you to add a pre-charge of air pressure to the tank. Oil accumulators deliver oil to the engine before starting (pre-oiling) to eliminate dry start scuffing of internal components, and discharge oil during low oil pressure surges to protect against engine damage during demanding racing conditions.

Oil accumulators are designed to collect pressurised oil from the engine and store it so it may be discharged later. At the time the engine is shut off and the tank valve closes, any oil pressure in the tank is held there. On engine start-up when the valve on the oil side is opened, the pressurised oil is released into the engine and lubricates the engine prior to start-up.

After the engine is started and the oil pump has taken over, oil is pumped back into the tank. This moves the piston back and pressurises the tank until it equalises with the engine's oil pressure. While driving, if the engine's oil pressure is interrupted for any reason, the tank releases its oil reserve again, keeping the engine lubricated until the engine's oil pressure comes back to normal. This release of oil could last from 15 to 60 seconds, depending on the size and speed of the engine. In racing or hard driving conditions, the tank will automatically fill and discharge when needed as you corner, accelerate and brake.

This system reduces the chance of oil surge. It costs more than a baffle or swinging sump set-up but can be used with a baffle plate or swinging sump. It provides a cheaper alternative to a dry sump system.

Dry sump

The dry sump system literally keeps the sump of the engine dry and allows the sump to be very shallow so the engine will fit lower in the chassis, giving a lower overall CG. The design of the external oil tank differs from a wet sump. The oil is still pumped into the engine at elevated pressure and then flows down to the engine's sump; instead of being held there, the oil is sucked away from the engine by one or more scavenger pumps, run by belts or gears from the crankshaft, usually at around half the crank speed.

In most dry sump designs, the oil reservoir is tall and narrow, and specially designed with internal baffles. The pump itself consists of at least two stages, but can have up to as many as five or six. With two stages, one is for scavenging while the second is a pressure stage. The three-stage

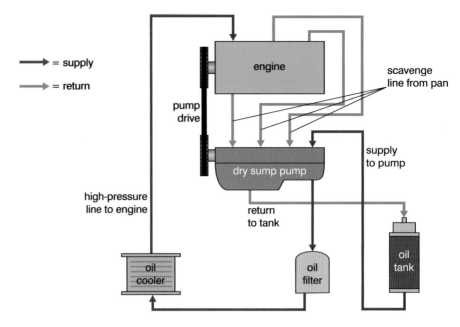

Figure 1.25 Dry sump system

dry sump pump has one pressure section and two scavenge sections. The four-stage pump has one pressure and three scavenge sections. The pressure section feeds oil to the block, while the scavenge sections pull oil from specially placed pickups in the dry sump oil pan. This system prevents oil slosh in the engine, reduces windage and increases horsepower. In some cases, a fifth stage is added to provide extra suction in the crankcase area.

So to summarise, the dry sump system allows a shorter sump pan, resulting in a lower CG for the engine/car, with aerodynamic and body packaging benefits:

- Control of oil pump speed is based on pulley size.
- There are options of two-, three-, four-, five- and six-stage systems – scavenge and pressure.
- A two-part tank can be used, allowing you to clean the tank internally.
- The tank removes air that is mixed with the oil, improving lubrication properties.
- Some tanks have pre-heaters to assist in heating the oil so that engine wear is minimised from cold starting.
- Vacuum or low pressure can be created in the crankcase, which will provide more power.
- The vacuum created gives a better ring seal, which can also provide more power.
- Cooling is improved due to a larger oil capacity.
- There is no oil for the crank to move through, which reduces friction.

The main disadvantage is that of cost and the fact that the system can rely on a belt to drive the gears for the pumps, which could be vulnerable.

Oil quality is important and should be replaced regularly in a race engine to ensure that its lubrication properties are not weakened by running at constantly high temperatures. Oil will run effectively at 90–105 °C; anything over this and you should consider fitting an oil cooler to control the temperature.

Oil pre-heaters are becoming more commonly used in all levels of racing, to heat the bulk of oil before starting and so reduce engine wear. An engine should not be driven hard until you see temperatures of at least 60 °C. However, keeping the oil around its main operating temperature of above 90 °C should see a small power gain due to the oil being thinner and causing fewer losses.

Oil pressure will normally sit at a pressure of around 60–80 psi when under load. During driving it should never drop below 20 psi.

1.8.2 Coolant

The coolant system should be a relatively easy part of the engine installation to get right. Thermal efficiency dictates that around only 35 per cent of the heat created in the cylinders is put to its original use – turning the crankshaft. A lot of the heat is absorbed into the surrounding material and the cooling system must take this heat away and dissipate it into the airstream.

An aluminium radiator should be used to provide the ultimate heat dissipation, while ducting should be used to guide fresh cool air in and ideally provide a place to dispose of the heat that is exiting the radiator. Keeping the fins straight and free of debris will assist with airflow, and a mesh should be fitted in front of the radiator if it could be exposed to a direct flow of debris.

Circulation can be controlled by the original water pump, but an electric water pump will provide greater control of flow rates and also allow you to run the water pump before and after the engine is switched off – reducing the chances of any hotspots building up in the cylinder head. When integrated with the engine ECU, an electric water pump can be controlled to operate to a specific temperature window. The ECU can also kill power to the water pump until a set temperature has been reached, allowing a quicker heat-up procedure. You should also see a small gain in power as the engine no longer has to drive the mechanical water pump.

An electric fan can also be used to assist with cooling in stationary and safety car periods, for example, but a leaf blower can be used for stationary periods in order to avoid adding unnecessary weight to the car. When an electric water pump is fitted, it can be programmed to alter its flow rate based on engine temperature (i.e. when it starts to get too hot, the pump will be able to circulate faster to boost water flow and heat dissipation).

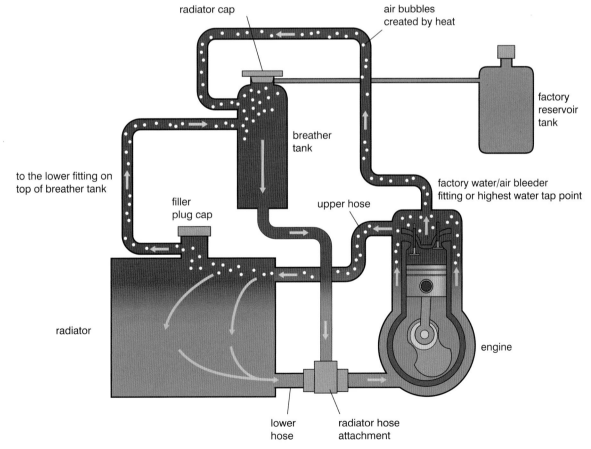

Figure 1.26 Coolant system

Air in the coolant system can result in severe damage to the system and having a system that allows air to easily escape is very important. Air bubbles or air locks will reduce the efficiency of heat transfer and by pressurising the system we allow the boiling point of the coolant to be raised, which reduces the chance of air bubbles forming. Normal water temperature is around 80–90 °C in a race engine, but this can rise when airflow across the radiator is reduced and the engine could easily reach 100 °C when switched off. A 1 bar (14.7 psi) pressure cap on the swirl pot can raise the boiling point to around 125 °C.

Remember, the temperature sensor only reads the temperature in its localised area and, if a mechanical water pump is present, the coolant that is sat next to the cylinders and is not being pumped around the system will be significantly hotter.

Some engine builders will retain the thermostat (sometimes with a few small holes in the plate to improve flow) or fit a restrictor on the exit of the engine in order to increase pressure in the block and head as this stops cavitation (formation of air bubbles) due to the water pump speeds. The water flow is slowed down in the engine and gives the coolant the chance to pull as much heat away from the engine as it can during a cycle.

The coolant fluid itself should be a mixture of water and an antifreeze that will provide freeze and corrosion protection. The proportion of antifreeze should be around 40–50 per cent of the total coolant volume, as less than this will not provide sufficient protection, while too much will affect the heat transfer capability.

1.9 Forced induction

Forced induction is a great way to find more power if the regulations for the class of the car allow it. Forced induction allows you to pack more air/fuel mixture into the engine, meaning that the volumetric efficiency can be well over 100 per cent.

Forced induction can be in the form of a supercharger or a turbo charger. The former being driven by the crankshaft via a belt or gear drive, while the latter utilises the expelled exhaust gases to propel itself.

The addition of this forced induction can be known as boost and is measured in bar or psi. As boost levels increase, compression ratio must be considered. There must be a balancing act between the two, as the more boost you run the more you effectively increase your compression ratio and, therefore, the more stress the engine is under. Compressing the air also creates heat so, based on boost levels, an efficient intercooler will be required to maximise the efficiency of the charge.

In the case of the turbocharger, there are a few things to consider:

- The size of the turbine (exhaust side of the turbo) can significantly affect the performance of the engine. A large turbine will create more boost at higher rpm but will be slow to spin up, which will not help low engine rpm, creating turbo lag. However, a small turbine will spool up quickly, giving a low-down power increase, but will restrict boost levels higher up in the rev range, so a balance is normally required between the two, based on application.
- You can buy turbos off the shelf, but hybrid turbos are more common in a racing environment as they are matched for the specific engine and type of use required.
- A turbo must have an oil feed to aid lubrication and cooling. When starting the engine, do not rev it up as it can take time for oil to properly lubricate the turbo bearings and revving the engine will increase the rpm of the turbo, possibly causing premature wear. Leaving the engine to fast idle after use allows adequate oil flow to the turbo to take away the heat and continue lubricating the turbo before the engine is switched off (when the engine is switched off, the oil flow is cut off so you do not want the turbo spinning fast when the engine is killed).
- A wastegate must be used to limit boost pressure in order to protect the engine.

Forced induction 35

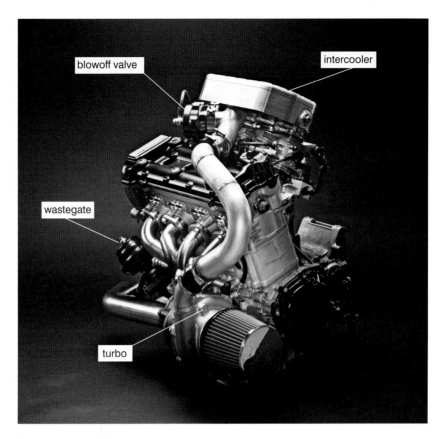

Figure 1.27 Turbocharged system

A supercharger is belt-driven by the crankshaft and so is more responsive as it does not have to rely on being spun up by the exhaust gases. The boost is generally limited by the drive ratio, which is determined by the diameter of the pulley on the crankshaft and the diameter of the pulley on the end of the supercharger unit.

Cooling is an important area to consider with any type of forced induction system due to the heat produced. A front-mounted intercooler is often used to provide fresh cool air to take away as much heat as possible from the charge entering the engine. Water injection is another way to ensure that the charge does not pre-ignite before the ignition point determined by the spark plug. This works by spraying an amount of water into the inlet tract under pressure in order to cool the inlet charge when it reaches a given maximum temperature, such as 40°C.

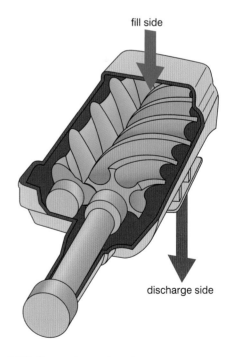

Figure 1.28 Supercharged system

1.10 Dynos and tuning

We have now discussed how power and torque can be found from an engine and how to build a package that works together harmoniously. To tie all of the parts together and to extract the best performance from the engine, you must spend time on the dyno in order to realise the engine's full potential. Every engine build is slightly different, so sharing maps from similar engines will not provide you with the optimum set-up.

There are two types of dyno – the rolling road and the engine dyno:

- Rolling road – a quick and easy way to find power output as you can simply drive the car on to the rollers and perform a power run. Rolling roads can be of an inertia type or a brake type (which includes a water brake or eddy current brake).
- Inertia dyno – this is a type of rolling road, which basically measures how quickly the car can accelerate a large drum, and then calculates torque and power. This is good for basic power runs to gain a print out.
- Brake-type rolling road – this is much more useful to the tuner, particularly the eddy current type, which is more user-friendly. The benefit of the brake-type rolling road is that it allows you to provide resistance to the rollers, which in turn allows you to hold the engine at specific rpm at a variety of engine loads, allowing you to finely tune the engine's fuelling and injection.
- Engine dyno – this is the one used most commonly by engine builders as it takes away any variables and reliably measures engine torque and power directly at the crankshaft. It is more time consuming to set up as the engine must be set up on to the rig but can allow for easier access to modify parts, such as inlet and exhaust systems. As

Figure 1.29 Rolling road

Figure 1.30 Engine dyno

the engine dyno also takes over the 'driving' part of the testing, it allows the tester to be able to change ignition and fuelling quickly to be able to find maximum performance.

You must be aware that different dynos can provide different power outputs, so once you find a reliable and skilful engine mapper, stick with them! Calibration, type of dyno and age of dyno can all be factors in what power output you end up getting. For promotional use, figures can also be 'adjusted' in order to try to gain more customers. Most dynos will have some form of weather station, which will look at atmospheric conditions, such as temperature, humidity, barometric pressure and vapour pressure, and will perform a correction calculation to your engine output, allowing you to gain more accurate comparisons of engine output throughout the year. The industry standard for measurements is now ISO rather than BSI, as it used to be. Most modern dyno cells now also have full air-conditioning cells, which reduce the correction calculations needed. They can match test conditions to those the engine will be performing in, such as very cold or very humid conditions.

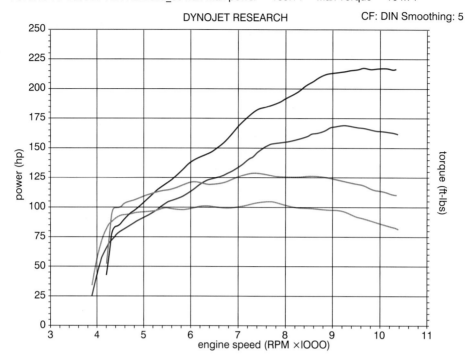

Figure 1.31 An example of a printout from a dyno test

CHAPTER 2

Transmission

This chapter will outline:

- Flywheel
- Clutch
- Gearbox
- Drive shaft and propshaft
- Differential

2.1 Flywheel

The basic purposes of a flywheel are:

- to store kinetic energy and pass this through the transmission
- to help carry the engine over between its fluctuating power strokes, helping it to run smoother
- to provide a ring gear around the circumference to allow the starter motor to engage
- to provide a flat friction surface for the clutch to engage and disengage with.

2.1.1 Calculating inertia

The flywheel is attached to the end of the crankshaft and requires energy to keep it spinning and to gain momentum. The **inertia** required to move this mass requires energy from the engine in order to accelerate it.

> **Inertia:** the resistance of any physical object to a change in its state of motion.

To measure the energy required, we can use the following equation:

> Energy required (J) = $I \times \left(\dfrac{\omega_2^2 - \omega_1^2}{2}\right)$
>
> Where:
>
> I = Moment of inertia (kg m^2) = Mass × Radius^2
>
> ω_1 = Initial angular velocity (rad^{-1})
>
> ω_2 = Final angular velocity (rad^{-1})
>
> **Note:** In these examples we are assuming all of the mass is distributed round the rim of the flywheel.

Worked example 1 shows how this can be applied to an example flywheel.

In this example, 30.4 hp is required to accelerate the flywheel in 3 seconds. The power must be generated by the engine and so would NOT be reaching the wheels of the vehicle during this stage. This is a considerable power loss simply due to the mass of the component. As such, for motorsport applications it is common practice to lighten a standard manufacturer flywheel, or fit a purpose-designed, lightweight flywheel of a different material. This reduces the moment of inertia of the part. A reduction of diameter would also reduce the inertia, but this is not as viable as it would require adaptation of clutch and starter arrangements to suit.

> **Worked example 1**
>
> Calculating horsepower required to accelerate a flywheel of diameter 300 mm and mass 13 kg, from an initial speed of 2000 rpm to a final speed of 7000 rpm in 3 seconds.
>
> I = 13 × 0.15^2 = 0.2925 kg m^2
>
> ω_1 = 2000 rpm = $\dfrac{2000 \times 2\pi}{60}$
>
> = 209.44 rad s^{-1} (2 d.p.)
>
> ω_2 = 7000 rpm = $\dfrac{7000 \times 2\pi}{60}$
>
> = 733.04 rad s^{-1} (2 d.p.)
>
> Therefore energy required:
>
> = 0.2925 × $\left(\dfrac{733.03829^2 - 209.43951^2}{2}\right)$
>
> = 0.2925 × $\left(\dfrac{493480.23}{2}\right)$ = 72,171.48 J
>
> Since:
>
> 746 J s^{-1} = 1 hp
>
> And:
>
> 72171.48 J over 3 seconds gives 24057.16 J s^{-1} (2 d.p.)
>
> Then:
>
> 24057.16 ÷ 746 = **32.25 hp** (2 d.p.)

Figure 2.1 Lightened flywheels

Figure 2.2 Lightened flywheel fitted to an engine

2.1.2 Types of flywheel

Typical flywheel options and example weights are as follows:

- Cast iron – most common for road-going production cars (13 kg)
- Lightened cast iron – the most simple upgrade from standard (8 kg)
- Billet steel – common upgrade for race-converted road cars (6 kg)
- Aluminium – used for purpose-built race cars (2 kg)

The effect of lighter flywheels

If the vehicle used in the previous equation is now fitted with a full race, aluminium flywheel with a mass of just 2 kg, then the horsepower required would be as shown in worked example 2 (note that in order to avoid repetition, the intermediate processes in this equation have been omitted).

Worked example 2

Calculating horsepower required to accelerate a flywheel of diameter 300 mm and mass 2 kg, from an initial speed of 2000 rpm to a final speed of 7000 rpm in 3 seconds.

$$I = 2 \times 0.15^2 = 0.045 \text{ kgm}^2$$

$$\text{Energy required} = 0.045 \times \left(\frac{493480.2}{2}\right)$$

$$= 11103.31 \text{ J (2 d.p.)}$$

Energy required per second

$$= 11103.31 \div 3 \text{ J s}^{-1}$$

$$= 3701.10 \text{ J s}^{-1}$$

$$= 4.96 \text{ hp}$$

This is a considerable reduction and, as a result, the throttle response and acceleration of this vehicle would be massively improved as this energy can now be transmitted through the rest of the transmission and into the road wheels.

The lighter flywheel will also have other benefits. For example, the lower mass on the end of the crankshaft will prolong the life of the crankshaft and its bearings, so it is generally considered that 'the lighter the better'.

However, there are some drawbacks. Since the moment of inertia is not only the reluctance of a component to begin to rotate but also its tendency to continue rotating, if the flywheel weight is reduced too much for a given application then it loses its momentum much more quickly and can cause an engine to stall or struggle in some demanding situations, such as when tackling steep gradients.

Note that, although the equations used in this section only refer to the flywheel, they can be applied to every rotating component within the engine and transmission system as they all require energy to accelerate.

When lightening flywheels, remember that the thinner the flywheel, the less heat capacity it will have (depending upon material choice). This means it could warp if overheated, so material around the clutch face should be retained. When removing material from a flywheel, it is important to leave an adequate amount in the centre and remove material from the outer edge to allow it to maintain strength and still reduce its moment of inertia.

2.2 Clutch

The clutch provides the transfer of torque from the engine to the gearbox and is used to engage and disengage the two systems.

Most race cars will use a multi-plate clutch in order to increase torque capacity and reduce mass of inertia.

Figure 2.3 Multi-plate clutch

2.2.1 Torque capacity

The torque capacity of the clutch is determined by its diameter, the clamp load of the pressure plate and the coefficient of friction of the face(s). Applying this, we can see:

$$T = N \times R \times F \times P$$

Where:

T = Torque capacity (Nm)

N = Number of friction surfaces

R = Radius of gyration (m)

F = Coefficient of friction

P = Pressure plate clamping force (N)

Example 1: A carbon multi-plate clutch used for racing car with 3500 N clamping force:

$$T = 8 \times 0.06 \times 0.5 \times 3500 = 840\,\text{Nm}$$

Example 2: A road car clutch with the same clamping force:

$$T = 1 \times 0.1 \times 0.3 \times 3500 = 105\,\text{Nm}$$

2.2.2 Clutch plate

The friction (or clutch) plate itself is sandwiched between the flywheel and the pressure plate.

There are various options when deciding upon a friction plate:

- Centre type – this can be sprung or unsprung. The sprung version will utilise radially positioned coil springs around the centre of the plate and also leaf-style springs between the two friction faces of the clutch plate. This minimises torsional vibration and shock from engagement. If the sprung set-up is not required, the unsprung centre will cause a harsher and more abrupt actuation but will reduce power loss and allow for a more direct engagement.
- Construction type – if you are not going to use a multi-plate clutch, you still have a vast range of clutches to choose from, including the paddle, segmented and full-face clutch types. The choice will be based on heat capacity, type of use and torque.

sprung clutch

unsprung clutch

Figure 2.4 Sprung vs. unsprung clutch

- Friction material fixing – the friction material can be riveted, bonded or sintered, with most systems using a sintered method. Riveting is the cheapest and probably the most reliable fixing over a long period of time, so is used in road car clutches. However, the rivets often do not cope well with heat and can damage the flywheel and pressure plate if material wear is not regularly checked.

- Friction material choice – this is the most important element, with the two main choices being a type of metallic clutch or a carbon clutch. Table 2.1 highlights the characteristics of different types of application.

organic

three-paddle

four-paddle

Figure 2.5 Motorsport clutches

Table 2.1 Characteristics of different types of friction material

Material	Characteristics
Metallic/semi-metallic	- Used in modern road cars - Similar to brake pad material - Metallic, sintered iron or copper based - Semi-metallic/organic, organic and metallic material
Organic	- Suited to road use - Can be used for light competition work - Relatively lightweight and compact - Smooth engagement and less prone to judder - Organic materials include glass and rubber
Sintered iron	- Material can cope with very high temperatures - Harsh on/off engagement - Special flywheel surface needed
Cerametallic	- Used in rally and some race applications - Often used as an upgrade to fast road cars with high torque levels (forced induction) - More resistance to higher power outputs when compared to metallic/semi-metallic clutches - Smooth engagement and less prone to judder
Kevlar	- Good durability and resistant to hard use - Similar engagement to organic - Can glaze in stop/go traffic until used hard again - High temperature operating range - If overheated, can become unstable and may distort without returning to original shape
Carbon/carbon	- One carbon face acts as a friction lining against the next carbon friction face - Lightweight - Low inertia - Torque capacity increases when hot - Long life expectancy - Low flywheel wear - Retains form at high temperatures - Expensive

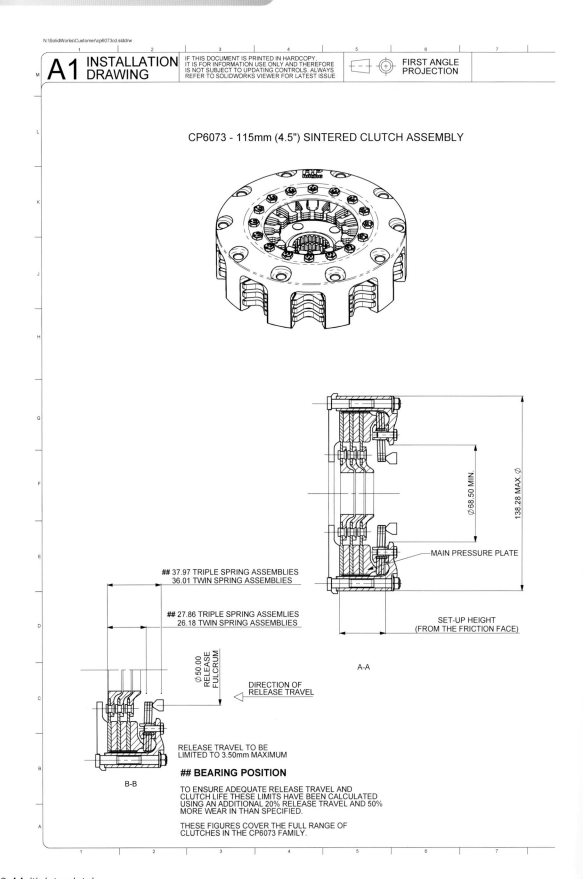

Figure 2.6 Multiplate clutch

CP6073 CLUTCH FAMILY

© AP Racing Ltd. 2004
AP Racing
Wheler Road
Coventry
CV3 4LB

Tel: +44 (0) 24 7663 9595 e-mail: engineering@apracing.co.uk
Fax: +44 (0) 24 7663 9559 Web site: http://www.apracing.com

Last Saved: jgovan on 27 October 2004 15:48:54

THIS DRAWING IS CONFIDENTIAL AND IS SUPPLIED ON THE EXPRESS CONDITION THAT IT SHALL NOT BE LOANED OR COPIED OR DISCLOSED TO ANY OTHER PERSON OR USED FOR ANY OTHER PURPOSE WITHOUT THE WRITTEN CONSENT OF AP RACING LTD.

Alterations

Issue No	Date & No.	Particulars	Zone	Initials
1	11/08/04 C2493	FIRST ISSUE OF RE-DRAW OF CP6074-1CD, INC. NEW PART NUMBERING SYSTEM.	#	#
2	08/10/04	WEIGHT WAS 2.78 ASSY. INERTIA WAS 0.0065 DP. INERTIA WAS 0.0013	#	JG
3	11/10/04 C2543	CP6074-22FM4 AND 023FM4 ADDED TO DP TABLE.	#	JG
4	27/10/04	SUH CORRECTED BY REMMOVING 2.5 FROM ALL FIGURES	#	JG

Specifications

MAXIMUM DYNAMIC TORQUE CAPACITY					
(Nm)	878	761	664	499	
(ft.lb)	647	561	490	386	
RELEASE LOAD					
Max. Peak Worn (N)	5500	5500	4700	3670	
At Travel (N)	4000	4000	3400	2670	
WEAR IN (See Note)	0.75	0.75	0.75	0.75	
Set Up Height New	33.52	33.69	33.39	31.87	
	32.38	32.41	32.11	30.63	
Set Up Height Worn - MAX	36.08	36.25	35.93	34.50	
Release Ratio	3.906	3.442	3.442	3.442	

Estimated Assembly Mass (Inc. Hub with Steel Main Pressure Plate) = 2.30 Kg
Estimated Assembly Inertia (Inc. Hub with Steel Main Pressure Plate) = 0.0055 Kgm²
Estimated Driven Plate and Hub Inertia = 0.00010 Kgm²

PERFORMANCE SUFFIX	DS	DE	SE	CE*	
For Reference					
Diaphragm Spring Rate	GLD	GLD	SLV	CRV	
Clutch Ratio	SHR	EHR	EHR	EHR	

* TWIN DIAPHRAGM SPRING.

MATERIAL SUFFIX	DRIVE PLATE MATERIAL	DRIVE PLATE THICKNESS
90	SINTERED	2.63mm

FLYWHEEL TYPE

	SUFFIX	COMMENTS
FLAT FLYWHEEL	FF	N/A
STEPPED FLYWHEEL	SF	FOR INSTALLATION DATA SEE SHEET 2

Sample AP Racing Part No. **CP6073-DS90-SF**

WEAR IN

THIS CLUTCH HAS BEEN DESIGNED FOR THE WEAR IN INDICATED ABOVE,
DRIVEN PLATE THICKNESS NEW: 2.63mm MIN
DRIVEN PLATE THICKNESS WORN: 2.34mm MIN

DRIVEN PLATES AVAILABLE WITH THE FOLLOWING SPLINE SIZES

SPLINE	PART No. STANDARD LENGTH (x 3)	PART No. INCREASED LENGTH (x 1)	PART No. INCREASED LENGTH (x 2)
1"X23T	CP5004-5FM4	CP6074-23FM4	CP6074-22FM4
7/8" x 20T	CP5004-6FM4		
1 5/32" x 26T	CP5004-8FM4	CP6074-19FM4	CP6074-18FM4
29.0 x10T	CP5004-7FM4		

SCALE 1:1 SHEET 1 OF 2
DRAWN: Jeremy Govan
APPROVED:
DERIVED FROM: cp6073-1cd (medusa)
TITLE: 4.5" (115mm) 3-PLATE SINTERED CLUTCH INSTALLATION DRAWING.
DRG NO. cp6073cd

DESCRIPTION

THE CP3759 SERIES CONCENTRIC SLAVE CYLINDER IS A LIGHTWEIGHT, HYDRAULICALLY SELF-CONTAINED UNIT WITH HIGH TEMPERATURE SEALS. ALL SEALING SURFACES ARE PROTECTED BY A HARD WEARING / LOW FRICTION COATING WHICH HELPS PROLONG SEAL LIFE.

ENSURE THAT THE UNIT IS INSTALLED IN THE ORIENTATIONS SHOWN, WITH THE BLEED PORT UPPERMOST. ALL FITTINGS INTENDED TO SEAT AT THE BOTTOM OF THE HYDRAULIC PORTS MUST HAVE AN INCLUDED ANGLE OF 90°

BODY MATERIAL: ALUMINIUM ALLOY (DIE CAST)
PISTON MATERIAL: ALUMINIUM ALLOY

EFFECTIVE AREA: 920 mm² (1.426 sq.")
MAXIMUM STROKE: REFER TO TABLE
MAXIMUM PRESSURE: 8.6 N/mm² (1250 psi)

HYDRAULIC PORTS: M10x1,0x11.5 MIN. FULL THREAD
HYDRAULIC FLUID: APR550 OR 600 BRAKE FLUID

HYDRAULIC FITTING KIT (STEEL ADAPTOR 7/16" '-4') - CP3759-6
HYDRAULIC FITTING KIT (STEEL ADAPTOR 3/8" '- 3') - CP3759-5

REPLACEMENT SEAL KIT - CP3759-3

ASSEMBLY No.	BRG FULCRUM Ø	REPLACEMENT BRG.
CP3759-50	50.00 (1.969")	CP3457-11
CP3759-54	54.00 (2.126")	CP3457-6
CP3759-38	38.00 (1.496")	CP3457-16

NOTE;
WHEN LARGE DIAMETER SPLINES ARE USED PLEASE CHECK CLEARANCE OF THE SPLINE IN THE SLAVE CYLINDER BODY.

LARGER SPLINES ALSO HAVE AN INCREASED DRIVEN PLATE HUB BOSS.
PLEASE CHECK THE CLEARANCE OF THE DRIVEN PLATE AT MAXIMUM SLAVE TRAVEL.
SEE THE CLUTCH ASSEMBLY INSTALLATION DRAWING FOR DETAILS.

FOR FURTHER INFORMATION PLEASE CONTACT AP RACING

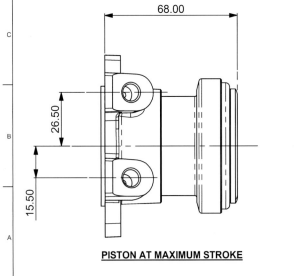

PISTON AT MAXIMUM STROKE

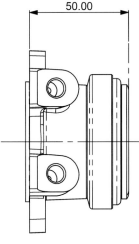

PISTON FULLY RETRACTED

Figure 2.7 Concentric slave cylinder

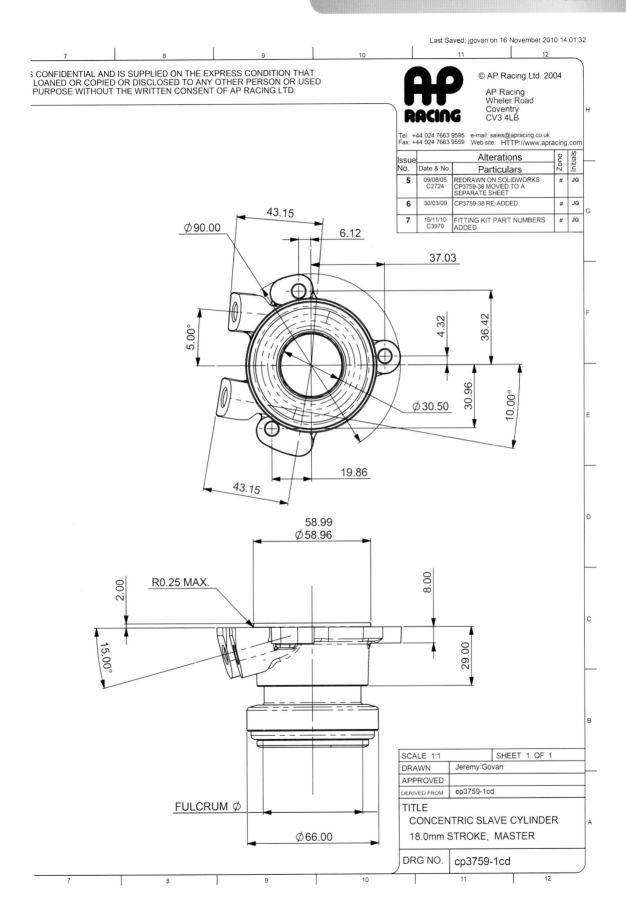

2.2.3 Pressure plate

The pressure plate sits over the top of the friction plate(s) and provides a clamping force, which presses firmly up against the friction plate in order to provide drive against the flywheel. Uprated pressure plates will provide a stronger clamping force in order to increase the torque capacity of the clutch assembly. When an increase in pressure is used, additional pressure will be required to disengage the clutch and, therefore, items such as the release bearing and the actuation system must be considered in order to overcome the fact that the pedal will become very stiff. Actuation methods can include cable or hydraulic and the feel can be altered via pedal travel, leverage and master/slave cylinder sizes. An external stop, often fitted in the pedal box, is needed in order to prevent the clutch from overthrowing and damaging the pressure plate springs.

There are two types of pressure plate actuation to disengage the clutch:

- Push type – this is a traditional set-up where the release bearing is pushed against the diaphragm spring fingers towards the flywheel and releases the clutch pressure.
- Pull type – this system has a release bearing fulcrum inside the clutch. It requires the spring fingers to be pulled away from the flywheel in order to disengage the clutch. The system tends to be more efficient in terms of clamping and release loads.

2.3 Gearbox

The gearbox is used to multiply the torque coming from the engine and to reduce engine speed to the required speed at the wheels of the car. It is put under tremendous strain, especially in a sprint race, where it will need to deal with hundreds of gearshifts, while endurance races require thousands of gearshifts. The gearbox often needs to be VERY compact when used in a transaxle scenario, as the suspension, exhaust, diffusers and wing all need to fit into this area of the car.

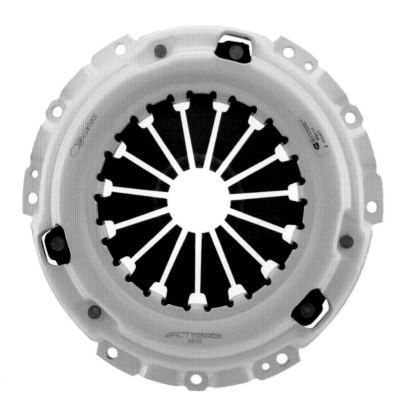

Figure 2.8 Pressure plate

Figure 2.9 Compact gearbox installation

2.3.1 Selecting the right gearbox

There are many options when it comes to selecting a gearbox:

- Gear type – helical (as in a road car) or straight cut
- Engagement type – synchromesh (as per a road car) or dog engagement
- Gear selection type – H-gate or sequential
- Actuation type – via rod or cable, hydraulics, compressed air or electronic (semi-automatic)

Other areas to consider will also be the torque requirements that the gearbox must be able to withstand and how many gears are required.

With mid-engined sports prototypes and single-seaters, the optimum transaxle system would be a semi-automatically operated, sequential, six-speed gearbox utilising dog engagement and straight cut gears.

Formula 1 cars have seven gears because they need such a varied range of speed within a relatively small rpm range and due to the high revs of the engine.

2.3.2 Inside the gearbox

A gearbox basically works by taking the drive from the centre of the clutch pressure plate, which is splined to the gearbox input shaft/clutch shaft.

The input shaft is in constant mesh with the layshaft. The 'driver' gears are splined to the layshaft and so turn at the same speed as the input shaft. These gears are then in constant mesh with the opposing 'driven' gears, which are fitted along the output shaft/pinion shaft.

The gears on the output shaft spin with those on the layshaft, but are fitted on roller bearings and so are not fixed to the output shaft. Instead, the

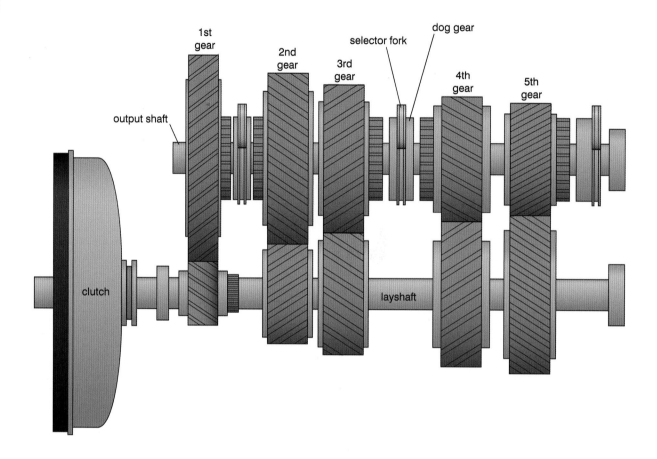

Figure 2.10 Gearbox layout

dog rings (controlled by the selector forks) are splined to the output shaft and have to engage with a gear in order to provide drive to the differential.

Note: some gearboxes do change gear on the layshaft instead of the output shaft, but these have never been competitive in racing.

2.3.3 Gear types

In motorsport the choice of gears is limited to helical and straight cut (or spur) gears. Helical gears are fitted as standard to most road cars and are capable of being used in light competition work. They have a large contact area due to the shape of their teeth and are quiet in operation. The downside is that they provide a sideways thrust, which causes a loss in engine power as it makes its way through the gearbox. Straight cut gears are noisier, as their teeth clash together, but the key advantage is that they provide a much smaller loss in power as the drive is directed correctly due to the shape and positioning of the teeth.

Gears are often made of a nickel chrome steel (or carbon steel such as EN55), but most Hewland gears are made of EN36. Carbon content determines the hardness of the gears so the carbon content may vary based on application. The gears are usually case hardened using a heat treatment, which provides strength and durability to the key contact areas of the gears.

Figure 2.11 Helical gear

Figure 2.12 Straight cut gear

2.3.4 Engagement type

The two main choices are synchromesh or dog engagement. Synchromesh systems are used in road cars. They allow two cones to build up friction and synchronise speeds before contact with the small dog teeth.

For a motorsport application, dog engagement is the most popular type of engagement as it is strong, simple, reliable and fast to operate. The selector fork (originally made of aluminium bronze, but now of steel) has a dog ring encapsulated within it which simply slides forward or backwards to mesh with a gear on the output shaft. A dog engagement gearbox does not need to use the clutch to go up or down the gears, only to pull away and stop. The less the clutch is used, the better the ability of the left foot to brake or act as a constant brace for braking and cornering.

The dog rings (also known as clutch rings) and the dogs on the side of the gears can be damaged easily, so gear changes must be carried out quickly in order to avoid the dogs clashing together and damaging the edges of the teeth. A rigid gear linkage must also be ensured to make the shift as direct as possible. The slower the shift, the higher the chance of dogs clashing together and becoming damaged.

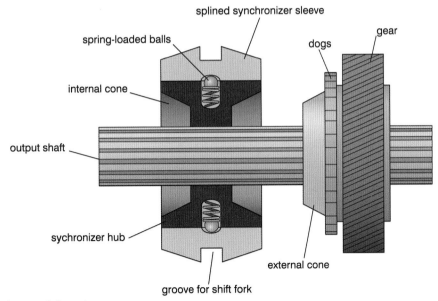

Figure 2.13 Synchromesh layout

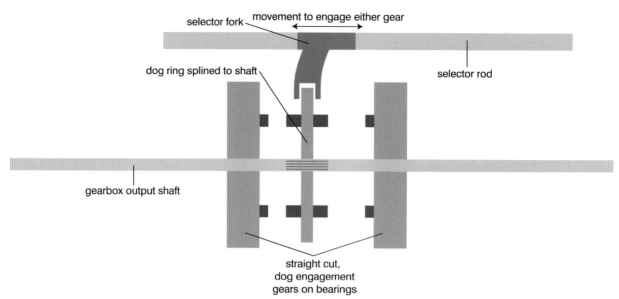

Figure 2.14 Dog engagement

2.3.5 Gear selection

The two types of gear selection are the traditional H-gate system and the sequential system (the racers' choice).

The H-gate has a linkage system that can operate one of three selector rods (sometimes four selector rods) that control an individual selector fork. Motorsport gearboxes often use a dog-leg first gear in order to enable the main useable gears (those after first gear) to be located in the main 'H', which makes it easy to shift within two selectors. As you rock the gear lever side to side, you are selecting a different selector rod, while moving the stick forwards or backwards moves the fork connected to the relevant rod from one gear to another.

In a Hewland-style gearbox without any gear-change assistance, an H-gate can often be as quick to use as a sequential system because the internal gearbox components are the same (except for the way in which the selectors are actually moved).

A sequential gearbox should still be the system of choice if budget permits. The sequential system simply removes the H-gate system, and often the selector rods as well, and places a ratcheting barrel above the selector forks, which is compact and direct. Each fork sits in a series of grooves or tracks around the barrel. As the barrel is rotated, the forks are moved along their tracks between the gears.

The sequential system provides the following benefits:

- Upgrade capability to a semi-automatic system
- More consistent shifting process (either forwards or backwards), giving the driver less to think about
- Consistent gearstick position that also requires a marginally smaller area for installation

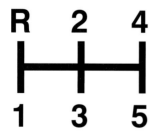

Figure 2.15 Dog-leg first gear layout, sprung against first and reverse

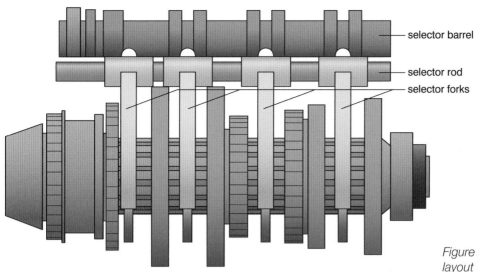

Figure 2.16 Sequential barrel layout for a motorcycle

2.3.6 Actuation type

In order to change the gears themselves, the link between the driver and the gearbox can be utilised with the following methods.

For an H-gate system, the choice is via a rod and spherical joint system or a cable system. The rod set-up tends to be most commonly used as it can be made rigid so that there is no flex or stretch, which provides a solid and instant gear change.

In a sequential system, the options tend to be more varied. They can be very expensive but provide more gains. A simple rod or cable system can still be utilised to give a normal mechanical linkage.

However, the options from there range from a simple flat shift system, which cuts the ignition on the way up the gearbox, which means you can still keep the throttle flat to the floor during upshifts. This is often sensed by a load sensor on the gear shift rod.

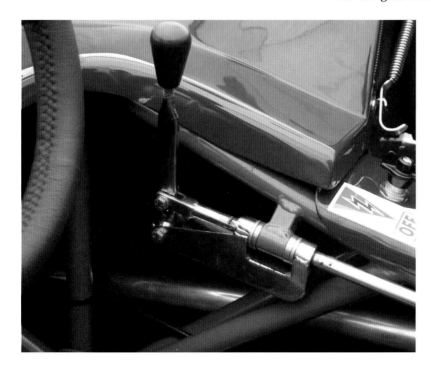

Figure 2.17 Rod linkage of a single-seater

Figure 2.18 Paddle shift on the back of a steering wheel

The next stage up is a semi-automatic system where the driver operates a set of electronic paddles behind the steering wheel or a pair of buttons on the steering wheel. From there, an electronic, hydraulic or compressed air system will actuate the gear selector arm and a throttle blipper can be fitted to the throttle bodies. The advantage of this system is that it will allow for flat upshifts and the driver can then match the revs automatically on the downshifts, meaning smooth and seamless up and downshifts at the press of a button. With this system, up and downshift times can be more reliable and carried out quicker. The driver can then begin to purely focus on braking, accelerating and steering, so lap times should fall.

2.3.7 Gear ratios

Gear ratios are vitally important to get right in terms of utilising engine performance and reducing lap times around a circuit.

Gear ratios are expressed as a decimal and are calculated by counting the amount of teeth on a pair of gears:

Gear ratio = Driven ÷ Driver

Note: remember the driven gear is on the output shaft and the driver gear is on the layshaft.

For example:

$$\frac{25}{26} = 0.96:1$$

$$\frac{34}{14} = 2.43:1$$

The ratio represents how many turns of the input shaft are made in relation to one turn of the output shaft.

Gear ratios should always stay as a matched pair. Once used, the gear teeth and dog teeth should be checked for damage, cracking and pitting.

A race car gearbox will have a closer set of ratios than a road car as it needs to stay within its powerband throughout the race, whereas a road car ratio set will also be designed with aspects such as economy and city driving in mind.

Selecting gear ratios

Selecting ratios can be a difficult task, particularly with the number available. A good starting point is to match the highest gear to the maximum speed that the car will reach at the circuit, which may be available from previous data or data from a similar formula. Bottom gear is usually selected and kept as the one that can give you the best start, whether rolling or standing.

You then need to stack the ratios between bottom gear and top gear appropriately, so that the engine pulls at its best from each corner to give maximum thrust on to the straight. You will normally expect the ratios to get closer together as they reach the higher gears and speeds, as the car needs to be able to overcome the rising amounts of aerodynamic drag. It is useful to have power and torque curves available when deciding on ratios once you have a saw tooth chart compiled of all of your available ratios. This will give you an accurate idea as to the rev drop to expect from each gear change and the rev range you need the engine to operate in. A car with a wide powerband is often more driveable and lap times from the average club driver will be more consistent as the car can be driven in a larger rev range. A car with a narrow powerband may be more difficult to drive consistently and it is difficult to set optimum ratios if it only works within a very small rev range.

Manufacturers such as Hewland provide a near endless list of gear ratios and the benefit of this type of transaxle gearbox is that it is quick and easy to remove the gear cluster from the casing and start to alter gear ratios. This is an excellent tuning tool to have at a race circuit, but knowing what ratios are available and how they will affect the car's performance is vital (hence using the power and torque curves with a saw tooth chart).

With most mid- to high-power/torque Hewland gear clusters, the layshaft has first gear machined on to it, meaning the whole shaft would need to be replaced if you wanted to change that ratio. Second gear is also often 'hubbed' in order to space the gears out.

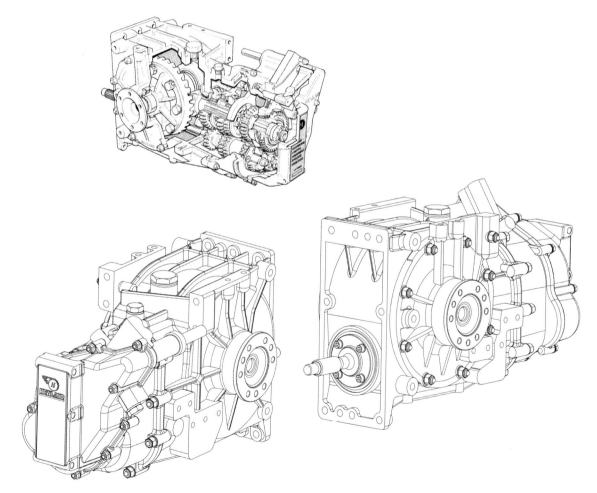

Figure 2.19 Hewland gearbox layout

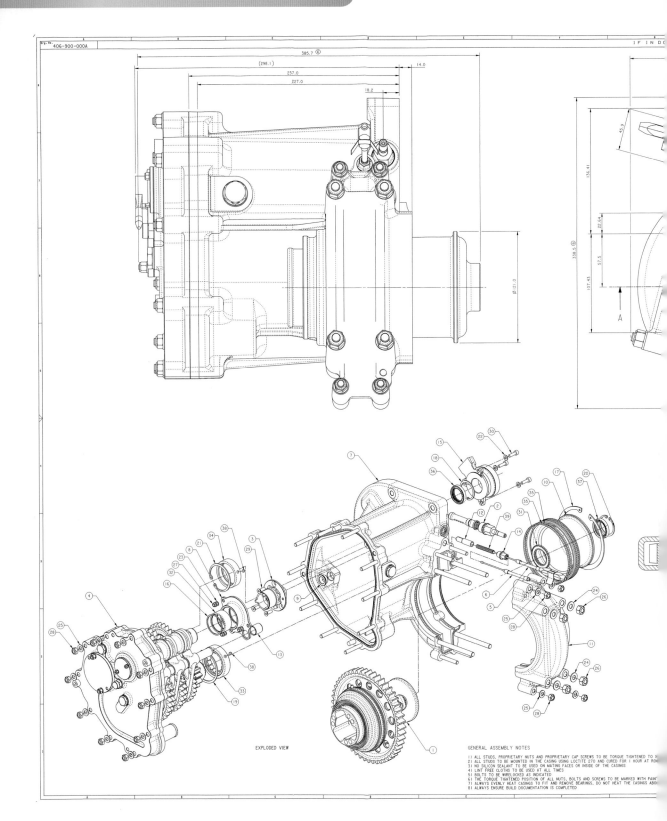

Figure 2.20 Gearbox assembly

Gearbox

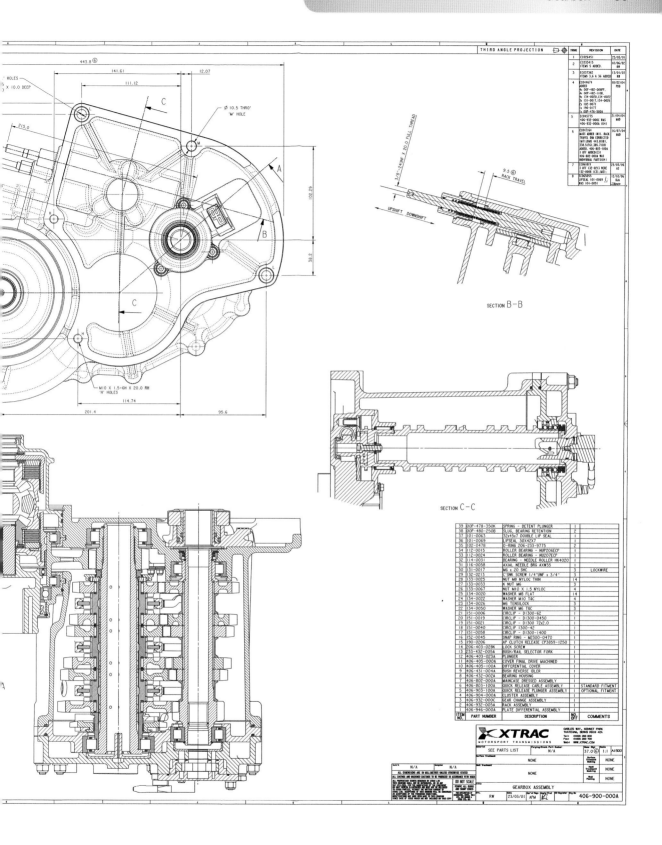

Calculating speeds using gear ratio

In order to calculate the speed of any race at a specific rev range, you can use the following equation:

$$\text{Speed (mph)} = \frac{\text{Tyre diameter (inches)} \times \text{Engine speed (rpm)}}{336 \times \text{Gear ratio} \times \text{Final drive ratio}}$$

Overall gear ratio can be determined by: Gear ratio × Final drive ratio

In order to make your saw tooth chart, calculate the speed in each gear ratio at two different rpm sites so that you have a straight line on your graph. Add all the other ratios you have to this chart. You will then have speed across the *x*-axis and rpm up the *y*-axis.

To compliment your saw tooth chart, it would be helpful to draw a vertical line to show the rpm at which maximum power, maximum torque and rev limit occur. With the help of the torque and power curve, you can then work out your shift point rpm and where the rev drop will push the revs back down to. By having the curve, you can aim to utilise the maximum area under the power curve in order to maximise acceleration. Data logging with a track map, wheel speed, rpm and throttle position can also be vital to determining the ultimate ratio set-up for each circuit.

Tractive effort curves can also be used to accurately consider the engine's performance regarding its pulling force in each gear compared with road speed. The equation for this is:

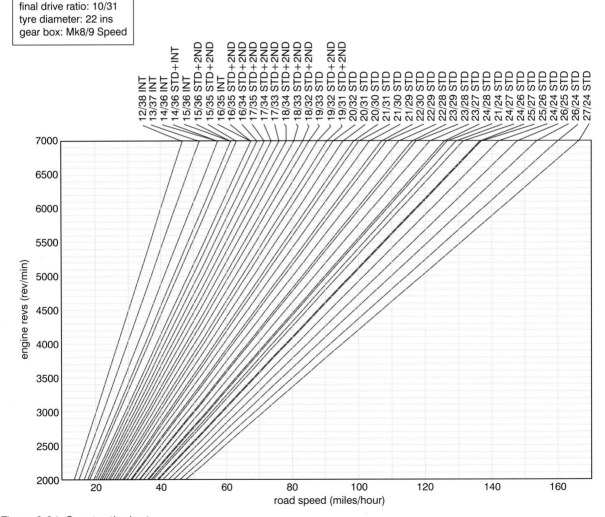

Figure 2.21 Saw tooth chart

> Tractive effort (N) =
>
> $$\frac{\text{Engine torque (Nm)} \times \text{Overall gear ratio}}{\text{Tyre rolling radius (m)}}$$

From this, you can plot the pulling force of the car along the *y*-axis and road speed along the *x*-axis to allow you to analyse the car's accelerating force across each gear and through the rev range by picking out torque figures at different rpm for each gear. You will then be able to see the forward thrust of the car in each gear and analyse where shift points, rev drops and which ratios should be used. You may notice, for example, that taking the car to the rev limiter in one gear is not worthwhile because the tractive effort in the next gear was greater at that same speed, resulting in wasted accelerating force.

A gearbox such as a Hewland will have a detailed manual that is normally downloadable from the manufacturer's website for you to keep handy for reference.

2.3.8 Torque limitation

A gearbox has a torque limitation just like a clutch. The diameter of the shafts and the sizes of the gears and bearings will determine the torque capacity of the gearbox. As an example, the Hewland Mk8/9 gearbox with its 1 inch diameter layshaft is rated to 150 lbs/ft (203 Nm) of torque, while the Mk5 gearbox has a 1.25 inch diameter layshaft and is rated to 180 lbs/ft (244 Nm) of torque.

If engine torque is known, the output torque of the gearbox can also be calculated.

For example:

> If gearbox efficiency was 85 per cent with 70 Nm of input torque from the engine and a first gear ratio of 5:1:
>
> Output torque = Input torque × Gear ratio × Efficiency
>
> $= 70 \times 5 \times \frac{85}{100}$
>
> $= 297.5$ Nm

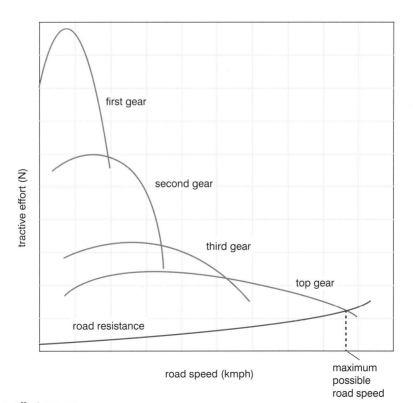

Figure 2.22 Tractive effort curves

2.3.9 Gear casings

The gear casings not only house the gearbox but also provide strength for the shafts to rotate and gear selection to occur. They must be rigid and yet lightweight, with casing material ranging from aluminium, magnesium and even carbon fibre. The gearbox, when used in a transaxle scenario, is often used as stressed member – the outer casing is used as part of the chassis and in most cases will have the suspension system and wing hung from various mounting points. Some gear casings can be over-engineered for certain applications and it is then possible to remove material from the casing in low-stress areas in order to reduce the weight.

2.3.10 Gear lubrication and cooling

Lubrication to gearboxes is normally splash fed, which relies on the oil in the gearbox being flung by the rotating gears to supply lubrication to all moving parts. Gear oil is 80 per cent coolant and 20 per cent lubricant, and you must always use new oil when changing gear ratios. Maximum temperature for most gearboxes is 115 °C.

More advanced systems utilise an oil pump and cooler just for the gear oil so that the temperature can be controlled and the oil scavenged from the bottom of the gearbox and fed back into the gearbox to guide the oil to critical lubrication areas.

Always refer to the gearbox manual when making any changes or adjustments and ensure you use the correct tools specified by the manufacturer.

2.4 Drive shaft and propshaft

The drive shafts (or half shafts) and propshafts are one of the most highly stressed parts of any race car, outside of the engine. They have to deal with constant torque fluctuations as the driver accelerates, brakes, changes gear, carries out a green flag lap and performs standing starts.

2.4.1 Drive shafts

The drive shafts have to deal with constant torsional loads and will need to be uprated along with the joints (constant velocity or tripod) from a standard car, as sticky tyres and an uprated engine will expose the half shafts as a weak link. The main stresses come from standing starts, changing up through the gears and hard acceleration out of slow-speed corners.

It is important to remember never to change the drive shafts to make them rotate in the opposite direction once they have been used, as this will put them at huge risk of failure after they take set due to the large difference between accelerating and braking forces.

The shafts must also contain no grooves or indents for the joint boots to sit in as these will cause high stress areas, and the grooves are not needed in order to seal up the joints. A metal tie wrap or just a plastic cable tie will be more than able to do the job.

The drive shafts must also be operable over the full range of the axle's bump and rebound travel. The joints at the ends of the shafts will have a certain level of built-in axial plunge to accept the vast changes in angle and length during suspension travel. If shafts are too long or short, bulking (through over-extension or compression) may occur, which will cause oversteer or wheel spin and failure of the shafts.

When specifying drive shafts, having them shot peened will remove any stressed areas and compact the metallic structure of the shafts, increasing its strength. Splines need to be cold rolled in order to reduce stresses that can occur during the manufacturing process.

Gun drilled drive shafts, which are hollowed out with a precision machining process, are lighter and also have better torsional resistance than those made from solid bars.

2.4.2 Propshafts

The propshaft is used to link the gearbox output shaft to the input of the differential on a front-engine, rear-wheel drive car. It must be balanced in order to avoid any potentially catastrophic vibrations, which can damage other areas of the driveline and engine. A centre bearing is used to allow for a change in angle and is usually a type of rubber bush that absorbs shocks and loads.

Materials for both drive shafts and propshafts vary; in most cases they are made of steel, but more exotic types can include aluminium or titanium and a mixture of composite materials, such as carbon fibre and glass fibre.

In the mid-engine, motorcycle-powered car world, a chain drive is often the most common form of system and the final drive is adjusted by altering either the front or, more commonly, the rear sprocket (which can be split into two pieces for ease of replacement).

Figure 2.23 Chain drive rear axle

Figure 2.24 Propshaft centre bearing

2.5 Differential

A differential is used in a vehicle to allow driven wheels to turn at different velocities from one input.

A differential is required because when a vehicle corners, the two driven wheels must travel at different speeds to stay in position on the vehicle. This can be seen by considering the distance the two wheels must travel: the radius of the track of the outer wheel is greater than that of the inner wheel and must travel further in the same time interval.

The percentage difference between the two wheels can be calculated by considering the difference between the circumferences of the two circles.

$$V_{diff} = \frac{2\pi R_0}{2\pi R_1} - 1 = \frac{R_0}{R_1} - 1$$

Where:

V_{diff} = Percentage difference in wheel speed (when multiplied by 100)

R_0 = Turn Radius, outer wheel

R_1 = Turn radius, inner wheel

For example:

For a car that has a track width of 1.5 m and is travelling a corner with a radius at its centre of 30 m:

$$V_{diff} = \frac{30 + \left(\frac{1.5}{2}\right)}{30 - \left(\frac{1.5}{2}\right)} - 1 = \frac{30.75}{29.25} - 1 = 0.051$$

This shows that there is a 5.1% difference in speeds between the inner and outer wheels at this time. Of course this changes as the radius of the corner does and is dependent upon the car's track staying the same and equal grip on both wheels, neither of which are constant in real life.

Most differentials used in motorsport are mechanical and rely on a few factors in order to change their characteristics of operation.

2.5.1 Differential types

There are three key types of differential:

- Open – fitted to most road cars, they are cheap, light and easy to make in very large production quantities but will direct torque to the wheel with the least amount of traction making cornering inefficient and adverse weather driving difficult.

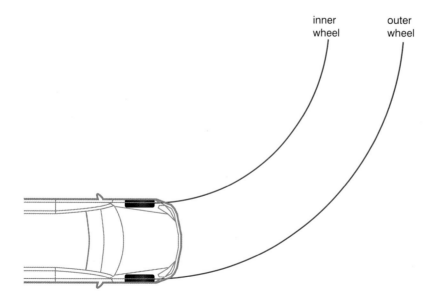

Figure 2.25 Differing corner radii

- Locked – Similar to the solid beam on a go-kart, the differential locks both drive shafts together. It has the benefit that both wheels will always have drive to them but scrub in corners is high and can cause problems with handling and tyre wear.
- Limited Slip – this provides the best of both the previous two by allowing the wheels to rotate at different speeds but also maintaining torque to the wheel with the most traction.

How do they work?

With most differentials, torque from the engine is delivered by the output shaft of the gearbox to a pinion gear. This can be directly from the output shaft (for example most front wheel drive and transaxle gearboxes) or via a propshaft to the differential assembly input. The pinion turns a crownwheel fixed to the differential and provides a final drive reduction due to the difference in tooth numbers between the two gears. The driven crownwheel turns the differential and this transmits power to the wheels through the connected drive shafts.

There are two common types of limited-slip differential (LSD) used in motorsport: the clutch/plate type and the automatic torque biasing helical gear type (ATB). The most obvious difference between the two is that a plate differential will always be attempting to return to a locked state where the two drive shafts are fixed together. An ATB differential never locks and operates by transferring torque away from the wheel with the least traction and towards the wheel with more traction.

The ATB helical gear differential contains a series of meshed pinions and sungears. The pinions are held in hardened pockets and connect together the outer sun gears which in turn are connected to the drive shafts. As the wheel speed on one side overtakes the other such as cornering or suddenly losing traction, the pinion gears start to turn. Energy will always take the easiest route and gears naturally try to force themselves apart. As the pinion gear is held in a pocket, the only direction of travel for the energy to escape through is along the length of the pinion gear towards the other side of the differential, thus transferring the torque away from the slipping wheel and towards the wheel with most grip giving a very efficient use of the available power. As the torque on the faster wheel reduces, the speed difference between the wheels will decrease and the differential will begin to balance the torque back across the axle. A typical bias ratio in an ATB differential is between 2:1 and 3:1 meaning that the maximum bias of power will be 2-3 times greater on the wheel with most grip than the wheel with least grip.

The advantages of an ATB differential are that it is totally automatic and requires no driver setup and no service intervals (as there are no wearing parts) and in operation, is the most efficient user of power from the engine and therefore ideally suited to fast road and track based motorsport.

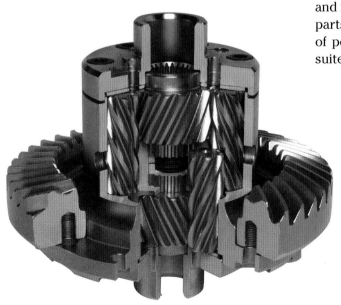

Figure 2.26 Automatic torque biasing (ATB) LSD

Drg No. F-2Q3-01		REMOVE ALL BURRS AND SHARP EDGES.	

Item No	Part No.	Title / Description	Qty
1	F-2Q3-02	FLANGED BODY	1
2	F-2Q3-03	END COVER - SMALL END	1
3	F-2Q3-04	SUN GEAR RH HELIX	1
4	F-2Q3-05	SUN GEAR LH HELIX	1
5	F-2Q2-09	FLANGE RETAINER	1
6	F-2Q2-10	FLANGE RETAINER	1
7	F-2Q2-11	CENTRE BLOCK	1
8	F-2Q3-12	END COVER - FLANGE END PLAIN TYPE.	1
9	F-2Q1-13	SPRING HOUSING	2
10	F-15Z1-07	PLANET PINION R.H. HELIX	6
11	F-15Z1-08	PLANET PINION L.H. HELIX	6
12	622	SOCKET HEAD CAPSCREW M10x1.5x20 LG	12
13	499	BELLEVILLE WASHER - $\varnothing 31.5 \times \varnothing 16.3 \times 1.25$ M	6
14	225	STEEL BALL - SUN GEARS	2

TITLE ASSEMBLY

SCALE 1:1 U.O.S. SHEET 2 OF 2

Drg. No. F-2Q3-01

A3(4)border

Quaife Design

Figure 2.27 Quaife ATB differential

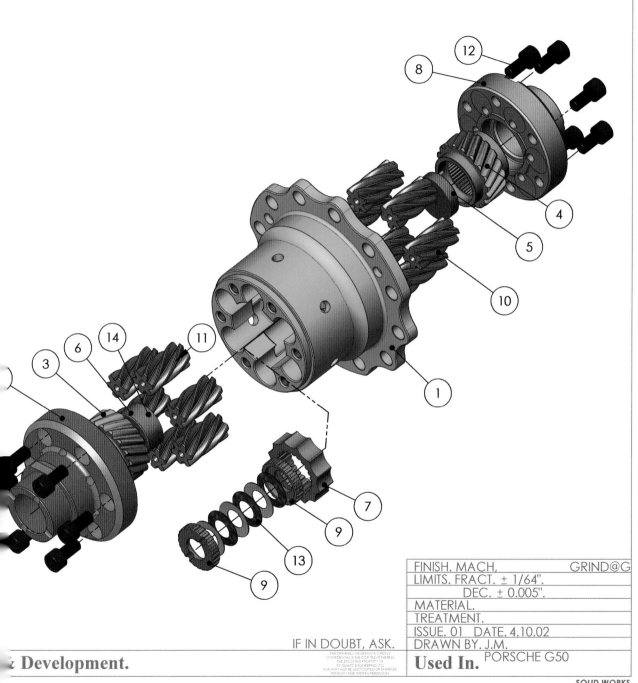

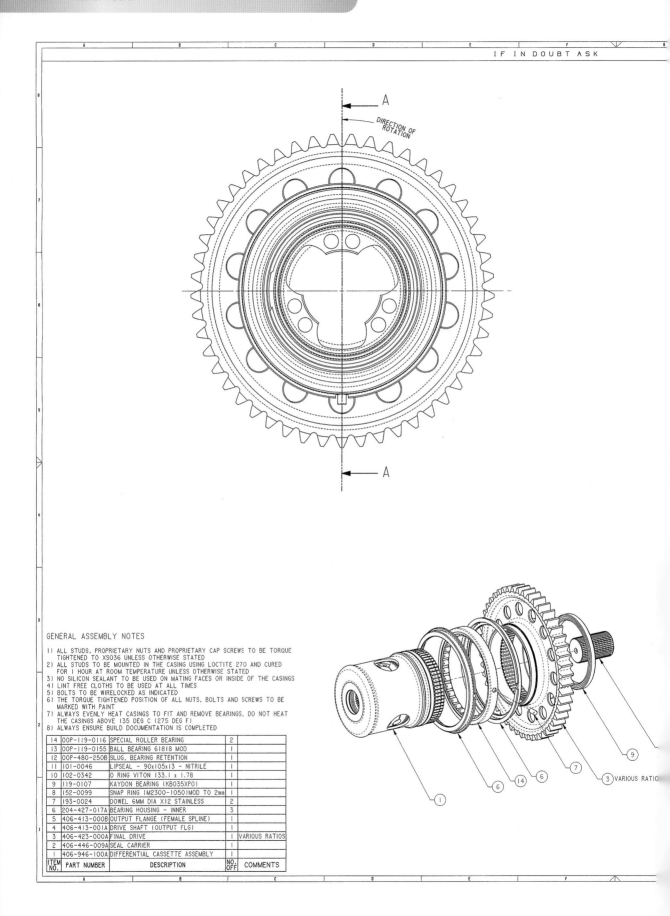

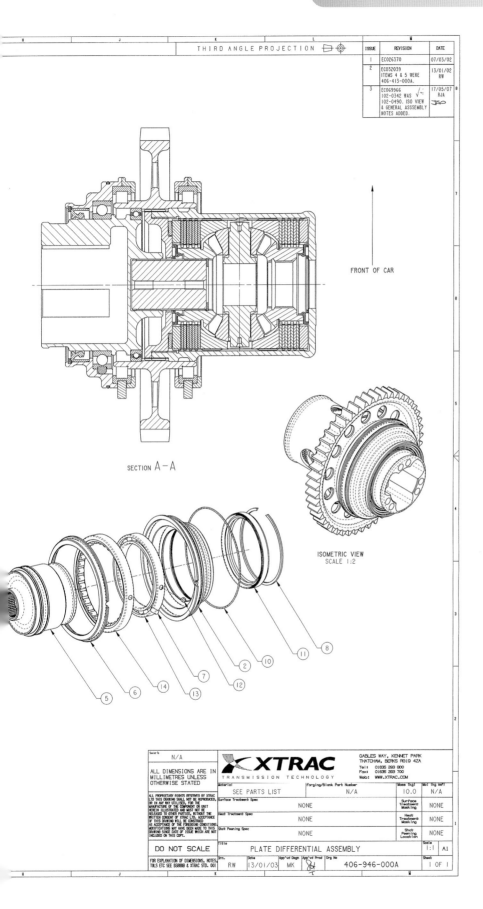

Figure 2.28 Clutch pack differential (plate type)

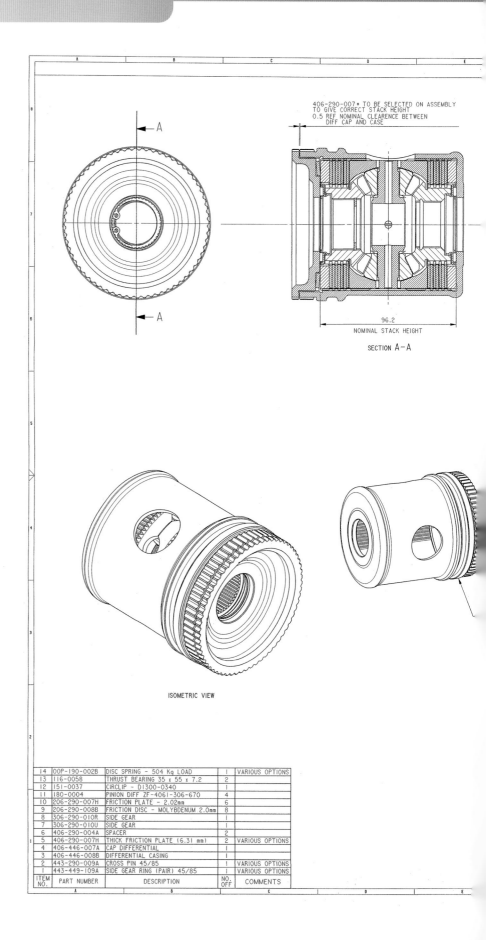

Figure 2.29 Clutch pack assembly

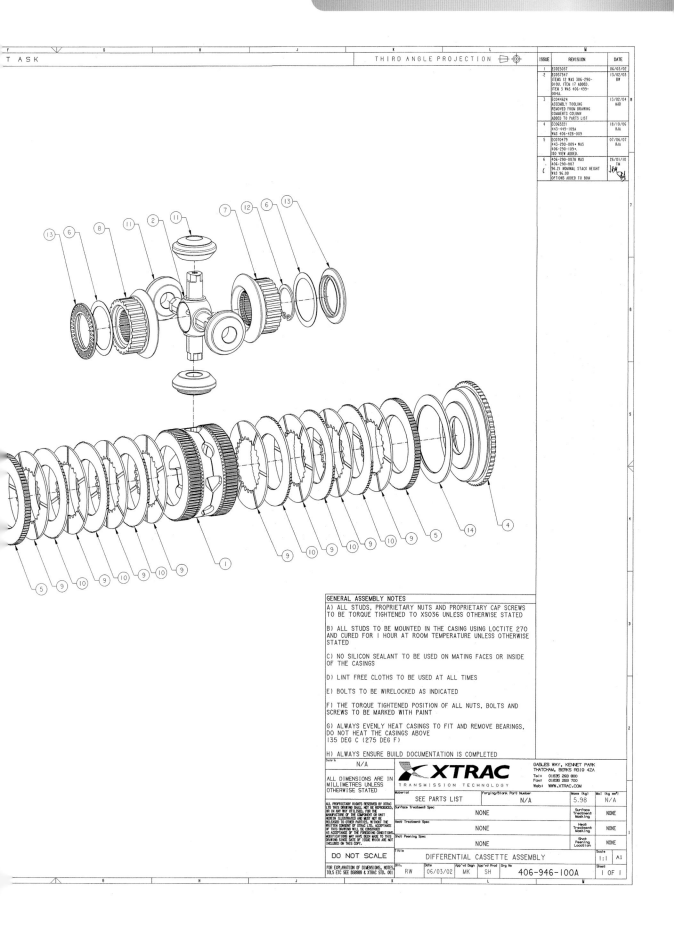

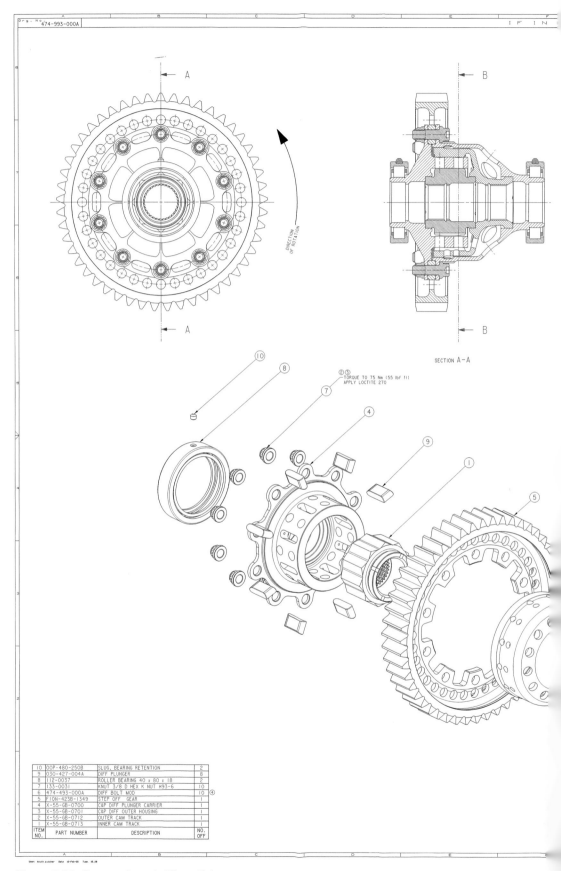

Figure 2.30 Cam and pawl differential

Differential

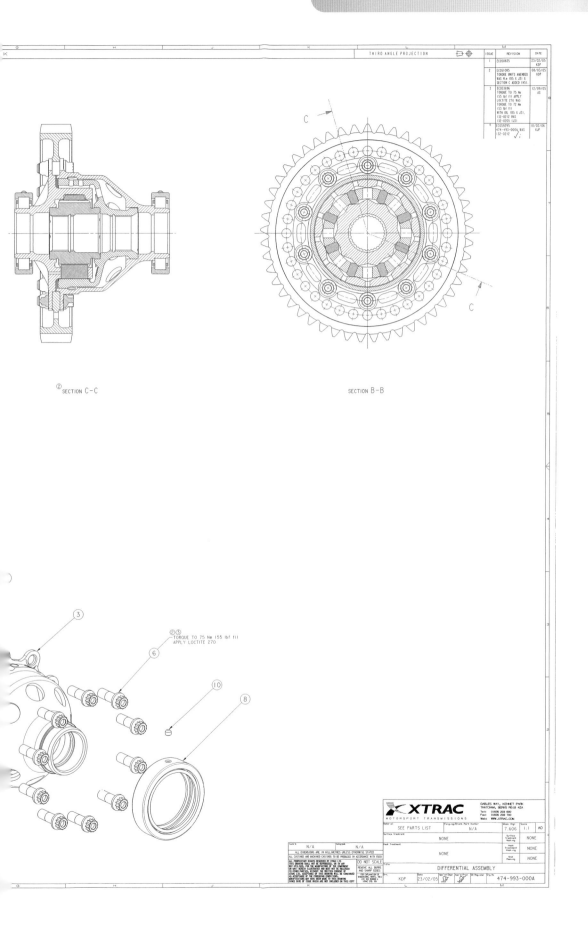

The disadvantage of an ATB is that if a wheel loses all grip, the differential will be transferring 3 × 0 torque to the other wheel therefore if a vehicle lifts a wheel (for example in rallying) all drive is lost until the wheel lands again. This also is true of a vehicle with a broken halfshaft

A plate style differential usually consists of a planet gears and side (sun) gears with a clutch pack housed at either end of the differential body. The planet gears are often located on cross pins to keep them from moving around inside the housing. The cross pins are sandwiched between reaction blocks which have ramps profiled into the sides. The clutch plates are positioned in such a way that one plate is splined to the main housing and the next plate is splined to the closest sun gear. The plates are set at a user defined preload, this is measured as the amount of force required to rotate or 'slip' (hence limited slip differential) the clutch plates against each other. In a straight line and with equal traction, the plates lock the two sides of the differential together. As grip on one wheel begins to lighten, the forces between the two wheels increases and when that force exceeds the preload force on the plates, they slip and the two sides of the differential begin to turn at different speeds (allowed by the planet gears). As the planet gears turn, they force the cross pins to ride up the ramps on the reaction rings, pushing the rings into the plate stack. The extra load now on the plates creates a higher preload and so the wheels move back towards a locked state. As the difference in speed between the wheels decreases, the cross pins run back down the ramps and the additional preload in the plates is reduced.

Adjusting the angle of the ramps will determine how fast the differential applies additional load to the plates and therefore how quickly the differential will lock up. Adjusting the preload of the plates will determine when the wheels will begin to slip (relative to each other). This gives the plate style differential the advantage of being configurable and users can adjust how the vehicle behaves at the entry to a corner (coasting/power off), during the corner (mid throttle) and on corner exit (power on).

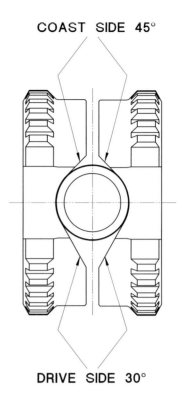

TYPICAL 30°/45° RAMPS SHOWN

Figure 2.31 Diff ramps

The advantage of the plate type differential is that if one wheel has no grip at all (such as in rallying) there will always be drive available to the other wheel.

The disadvantages of a plate differential are that this type of differential is based on friction and the plates wear meaning they need adjustment and eventually replacement- not an easy job in most front wheel drive gearboxes.

Cam and pawl type differentials are another type that are used in racing, but are maybe not as well known. They have an inner and outer cam track and are driven by the plunger carrier (attached to the crownwheel), which normally has eight plungers. The plungers will move in and out of the cam tracks acting as a free differential, until enough load has been applied to jam the plungers between the two tops of the cams and drive both wheels.

CHAPTER 3

Chassis

This chapter will outline:

- Tyres
- Suspension
- Steering
- Brakes

3.1 Tyres

Being the only contact that a car has with the race surface, the tyres are the most important part of any race car and deserve a great deal of attention and care. The design of the car must be optimised in order to maximise the performance of the tyres.

3.1.1 Tyre basics

The tyres have to deal with a variation of loadings throughout a race. They must initially provide grip, which must then extend to acceleration, braking and cornering forces, when pushed to the limit. They will be expected to do two things at once: brake and corner, and accelerate and corner. As well as this they must provide feedback so that the driver can work the tyres to their maximum grip levels at all times.

The characteristics of a tyre are determined by its compound and physical structure. The compound determines how hard or soft the tyre is, which affects grip level and tyre duration.

Tyre construction and structure

Cross-ply tyres have their plies angled at 45° from the tyre's rotation and cross over one another. They generally run with less static camber as the plies are shared with the sidewall and tread, so these distort together. Due to the elastic flex in a cross-ply, you will find that it will give more feedback and be more progressive, in terms of grip levels, than other types of tyre and so works across a larger slip angle range.

Radial-ply tyres are more commonly used and have plies that run at 90° from the tyre's rotation. As they have more plies running around the tread, the sidewall flexes, which requires more static camber to be used. They often provide more peak grip but are more likely to break away with less feedback. Due to the lack of flex in the tread pattern, the transient response of the tyre is also often superior to a cross-ply tyre.

Slick tyres are used in dry conditions and contain no grooves or cuts in the tread, which gives maximum contact patch and, therefore, grip.

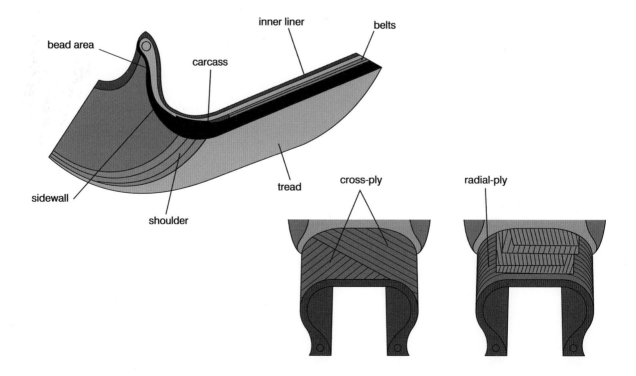

Figure 3.1 Cross-ply vs. radial-ply tyres (showing general tyre structure)

Wet tyres have a large amount of grooves and cuts in them to help channel the water away from the contact patch to avoid aquaplaning. The wet tyre often uses a softer compound as the water cools the tyre quickly so a softer tyre is needed to keep it at the optimum working temperature. Intermediates are somewhere between the slick and wet tyre in terms of tread pattern and compound.

slick

wet

intermediate

Figure 3.2 Slicks, wets and intermediates

The importance of vertical load and slip angle

In order to enable optimum grip levels, it is important to maximise the contact patch of the tyre to the track surface. The contact patch is the area of the tyre that is in contact with the ground. The larger the contact patch, the more grip there is.

The contact patch is affected by the vertical load (produced by downforce and vehicle weight), tyre pressure, camber angle and slip angle.

The more vertical load produced, the more grip the tyres will have. Vertical load is measured in newtons (N) or kilonewtons (kN) and is produced by aerodynamic load and weight transfer.

The coefficient of friction (Cf or μ) of the tyre is determined by:

$$\mu = \frac{\text{Side load}}{\text{Vertical load}}$$

Obviously, the higher the number, the more grip is available; however, with the addition of vertical load, μ reduces. Cornering force rises at a greater rate with vertical load than it reduces with μ (Figure 3.3).

Tyres also operate at an optimum slip angle, that is, the angle between the direction of the tyre's travel and the angle of the tyre's contact patch. The aim is that the car is prepared and driven so that the tyres are used at the optimum slip angle, providing a peak grip level and producing a fast lap time.

The tyre pressure not only alters the size of the contact patch but also the characteristics of the tyres in terms of their spring and damping rate – a higher pressure increases spring rate and reduces damping of the tyre, while a lower pressure reduces spring rate and increases damping.

A negative camber angle can also allow the car to produce a higher cornering force, but there must be a balance between optimum camber for cornering and the limitations in braking and corner exit that may occur from a reduced contact patch when the car is not in a roll state.

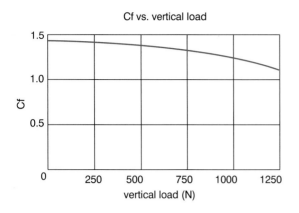

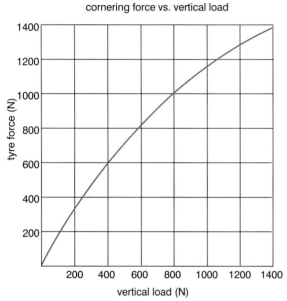

Figure 3.3 Effects of varying vertical load on Cf and tyre force

3.1.2 Maximising tyre performance

Tyre manufacturers with a designated motorsport department will often supply tyre data for their products, including recommended hot pressures and effective tyre spring rates at differing pressures. A tyre's spring and damping rate is based on its construction and pressure when hot, which affects the whole suspension spring and damping rate and so alters the performance of the car when switching between different tyre manufacturers or using different tyre pressures.

A G-plot of tyre performance can be created to measure driving and car performance by looking at cornering, braking and accelerating forces, as well as combining them together (Figure 3.5). Different tyres can produce a different G-plot. The larger the G-plot, the more grip that is available, so the quicker the car can go. A race tyre should be able to generate almost equal force in acceleration (power permitting), braking and cornering. Peak grip in braking and cornering can slightly differ as the traction circle can often be an oval shape. When combined, as shown by the purple line in Figure 3.5, they will not be able to produce maximum longitudinal and lateral grip; so when cornering the driver should aim to be as close as possible to the outer traction circle, maximising the whole cornering process at entry, apex and exit.

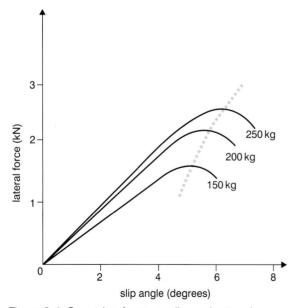

Figure 3.4 Cornering force vs. slip angle at various vertical loads

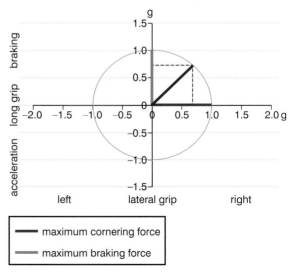

Figure 3.5 Tyre G-plot

Setting cold (static) pressures and controlling them will allow optimisation of the hot (dynamic) tyre pressure and, therefore, control the performance of the tyre. To set cold pressures, consider:

- circuit characteristics (e.g. type of corners, circuit direction, length of straights and type of surface)
- target hot pressures
- ambient and track conditions, such as temperature and track state (e.g. green or rubbered in).

Pressure

As the tyres are worked, they produce heat and, as a result, the tyre pressure increases until it plateaus according to the conditions the tyre is being used in (ambient, track temperatures, etc.). Normal air contains varying moisture levels so has different expansion and contraction rates, which is why teams like to use a stable, inert gas such as nitrogen. To gain maximum benefit from using nitrogen, you must purge the tyres before final inflation in order to ensure all air and moisture is removed.

When checking tyre pressures, you must remember that the readings you take are only as accurate as the pressure gauge used, so ensure it is calibrated. It is worth checking pressures twice prior to each session.

Table 3.1 Tyre pressures: cold and hot and corrected

Initial cold pressures (psi)	
15	15
16	16
Hot pressures (target pressure of 22 psi)	
23	22
24	21
New cold pressures for the next session to aim for 22 psi all round	
14 (−1)	15
14 (−2)	17 (+1)

Tyres should be marked with position, direction and set number in order to make them easily identifiable.

Figure 3.6 Tyre identification system

Temperature

Tyre temperature measurements determine the optimum use of the tyre contact patch. Temperatures can be measured in a variety of ways but the two common methods are to use an infrared sensor to monitor tyre surface temperature dynamically during the race, or to use a pyrometer with a tyre probe to measure bulk temperature immediately after the race (starting with the hottest tyre and working down to the coldest). The temperature needs to be measured across the tyre in three places: inside, middle and outer.

You would normally expect the temperature of the tyre to be marginally hotter on the inside of the tyre due to the amount of negative camber used on the front and rear wheels; this puts more load on the inner part of the tyre in a straight line and reduces the load on the centre and outer parts. The front wheels may use more negative camber and so we would expect a higher difference from inner to outer.

Fronts would be expected to have a difference of around 10–15 °C, while the rears may have a difference of 5–10 °C, due to running less camber at the rear based on traction issues that can occur from running too much camber.

Table 3.2 Tyre temperature readings and evaluation

Temperatures (°C) (outside/middle/inside)	Set-up conclusion	Balance conclusion
82 88 94	Good	Good
76 88 99	Too much camber	Tyre not optimised
86 88 90	Not enough camber	Tyre not optimised
86 85 87	Tyre pressure too low	Tyre not optimised
82 92 90	Tyre pressure too high	Tyre not optimised

You can also monitor the difference between the average front and rear tyre temperatures to analyse how the car is behaving. Differing temperatures (front to rear) could indicate problems, such as traction issues, poor chassis set-up or incorrect compound selection.

Tyre temperatures are generated by working the tyre through:

- brakes radiating heat
- pre-event heating techniques, such as tyre ovens and blankets
- sidewall and tread deflection, which creates strain energy
- tyre slip (longitudinal or lateral), which creates friction energy.

Generating tyre temperature and pressure is important. A green flag (warm-up) lap is allowed for cars with slick tyres and this allows the drivers to generate higher temperatures in the tyres. Performance over the initial few laps of a race is vital as this is often where most places can be won or lost.

Methods to warm the tyres by driving can include the following:

- Steering/weaving – this produces heat in the front tyres by creating lateral strain energy. Minimal heat is produced in the rear tyres.
- Braking – this allows the heat created in the brakes to be radiated into the wheel/tyre assembly. It produces longitudinal slip at the same time and can help to heat both front and rear tyres. The ratio of this depends on the layout of the car.
- Acceleration – hard acceleration creates longitudinal wheel slip and will create heat in the driven wheels.

A mixture of the above on a warm-up lap and during a safety car period will help build tyre temperature.

Tyre temperature is also dependent upon ambient and track conditions, as discussed with tyre pressure.

3.1.3 Tyre issues and looking after tyres

Tyre degradation is when a tyre wears out during a race. All tyres degrade but at different rates, depending on factors such as compound, vertical load and track temperatures. The more worrying aspects of tyre degradation are covered on the next two pages.

Graining

Graining occurs when a large temperature difference between the surface and bulk of the tyre causes a shear failure. This often happens when the tyres are pushed too hard over the first few laps when tyre temperature has not been built up correctly, when the track/ambient temperatures are very cold or when the selected compound is too soft. This is a scenario that needs to be avoided.

Figure 3.7 Tyre graining

Blistering

This occurs when the surface temperature of the tyre is exceeded so that the tyre tread heats up, softens and peels away from the tyre plies. This can occur from excessive wheel lock-up, too much camber applying too much load to the tyre shoulder, or when the wrong tyre type is selected (e.g. a wet tyre on a quickly drying circuit or a wrong compound).

Figure 3.8 Tyre blistering

Tyre hardness

A tyre's lifespan depends on a number of factors, but heat cycles are the main factor in changing a tyre's compound. A heat cycle is a period where the tyre is heated and then cooled. This process also cures the rubber and, therefore, the more heat cycles a tyre experiences, the harder the tyre becomes.

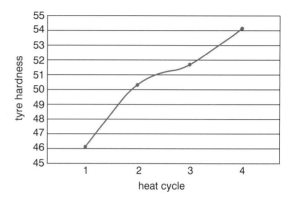

Figure 3.9 Tyre hardness vs. heat cycles

A tyre durometer can be used to measure the shore hardness of the tyres and, although in the higher levels of motorsport tyres are replaced frequently over a weekend, this is not always possible for a club racer. A rule of thumb is that you have normally got the best out of the tyre after six heat cycles, as after that you will find a slight and gradual decline in the tyre's grip levels. Therefore, the more often you can replace your tyres, the quicker and more consistent the car will be.

Figure 3.10 Tyre durometer

Tread depth

Tread depth needs to be monitored if you use the tyres over a long period. Wet tyres will be affected the most as the passages for the water to be dispersed are reduced in their volume capacity. Slick tyres have tread depth indicators in the form of holes; however, slicks can often give limited warning before the belt plies become visible.

Storage

Tyres also need to be stored correctly to enhance their lifespan. Keeping tyres at room temperature in tyre bags (or black bin bags) will shelter them from UV light (fluorescent lighting tubes in particular) and stop the shore hardness creeping up.

Cleaning tyres

At the end of an event, you may find on the cool-down lap that the tyres pick up rubber from the track and other debris. As the tyres do not reach high loadings on the cool-down lap, they come into the pits covered in rubber lumps that have been discarded from other tyres. Some people choose to leave the rubber and debris on the tyres and will try to remove it on an aggressive warm-up lap. Others will clean the tyres between sessions, either with a heat gun and scraper, or by using a wood plane and soapy water.

Bedding in

New tyres, like most race car parts, need to be bedded in. When tyres are made, a release agent is added around the outside to help remove the tyre from its mould. This slippery coating needs to be worn away from the tyre before it will grip; this should only take one or two laps to scrub the tyres in. After this, the tyre will be at its stickiest, which provides a great advantage during qualifying, while a lightly scrubbed set can be used during the race. Depending upon the car, tyre type and circuit, it is usual to see lap-time savings of half to one second per lap with new tyres.

3.1.4 Wheels

Wheels (or rims) have a key role to play. They support the tyre around its bead, while allowing it remain inflated. As they form part of the unsprung mass of the car, they must be lightweight but also stiff in order to avoid flexing under heavy loads. Motorsport wheels are usually made of aluminium, magnesium or, in some cases, a composite such as carbon fibre or Kevlar. Some formulas, such as FF1600, specify that all wheels must be made of steel. The width of the wheel is important as this provides the seat for the tyre and the tyre width must be matched to the wheel width. According to the Avon Tyres Motorsport website, a 230mm wide slick tyre has an optimum wheel width of 10 inches, although 9–11 inches is suitable.

Wheels can be one piece and are often cast, while some manufacturers will use three-piece wheels (with an inner and outer rim fixed to a centre – billet or cast), which can often provide greater freedom to gain desired rim widths and offsets and allow for alteration and ease of repair in the future.

Attachment is often in the form of either a conventional four- or five-stud bolt pattern (with wheel studs and nuts preferred over wheel bolts) or with a single-centre lock nut that locates the wheel to the hub with a series of drive pegs and holes. Of course, the centre lock set-up allows for a much quicker changeover from one set of wheels/tyres to another.

Tyre valves can be vulnerable when racing is close and competitive, so short stubby valves should be used and valve caps should always be used to protect the valve from being fouled by debris. They should also be balanced between each event in order to prevent any unwanted vibration, and the weights should be covered in foil tape as per regulations.

3.2 Suspension

The role of the suspension system is to ensure that the tyre stays in contact with the road surface while maintaining a large contact patch and absorbing bumps and undulations from the road. It also offers tunability to control how the chassis reacts in a multitude of conditions.

Although still very mechanical, the suspension of the car is a critical system that connects the sprung mass to the unsprung mass and, most importantly, controls the tyres' contact with the road.

3.2.1 Sprung and unsprung mass

Sprung mass is the weight of the vehicle that is supported by its suspension springs, such as the chassis, engine, driver and gearbox. The heavier the sprung mass, the stiffer the spring must be to keep the car off of the ground.

Unsprung mass is the weight that is not supported by the springs, such as the wheels, tyres, brakes and uprights. As this is effectively uncontrolled, the lighter it is, the better the contact between the tyre and road surface, particularly during transitions and when tackling bumps and undulations.

3.2.2 Springs

Springs can change all aspects of a car's handling, including its ride, roll and pitch rate. With adjustment at each axle, the car's overall balance can also be changed to control understeer and oversteer. They are also used to store mechanical energy from forces produced at the tyre.

Figure 3.11 shows pitch, roll and yaw, which are known as the three degrees of freedom. Roll occurs laterally during cornering; yaw occurs with understeer and oversteer; and pitch occurs longitudinally during acceleration and braking.

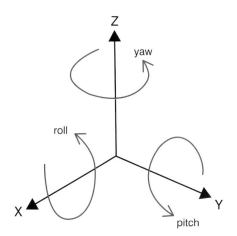

Figure 3.11 The three degrees of freedom

Ride rate is essentially how the car reacts to the road surface when it encounters bumps and undulations. Minimising pitch and roll angles allows the car to not only stay closer to its static geometry, but also to assist in keeping the aerodynamic balance consistent during accelerating, braking and cornering. The springs also control heave caused by the aerodynamic downforce pushing the car into the ground (ride height control).

Stiffer springs can control more force when coupled with a damper, but generally do not follow the road as well as a softer spring.

When fitted in series (inline) the initial spring rate (K) can be calculated using the following equation:

$$K = \frac{k_1 \times k_2}{k_1 + k_2}$$

k_1 and k_2 are substituted for the spring(s) (main and tender if using two) in use.

When fitted in parallel, for example in specially built off-roaders, the equation changes to:

$$K = k_1 + k_2$$

Coil spring types

Compressive coil springs, which are the most commonly used in race cars, have their stiffness determined by the thickness of the sprung steel wire coils and by how many active coils are in use.

The springs can be grouped into the following categories:

- Main (linear and progressive) (Figure 3.12) – this is the only spring that can be used alone. A linear spring has the same spring rate as it is compressed through its travel; it has an inactive coil at either end where it has been closed. A progressive spring has a spring rate that increases as it compresses. This is achieved by varying the amount of active coils; as the spring compresses, some of the coils bottom out.

- Tender (Figure 3.12) – this spring needs to be used in conjunction with a main spring and fitted in series; an aluminium joining collar is necessary to keep them inline. This set-up provides a softer initial rate as both springs are compressed together and can give a softer ride rate. Once the tender spring has completely closed, the main spring rate controls downforce, pitch and roll. Tender springs are available in a range of linear and progressive spring rates.

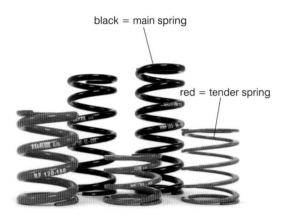

Figure 3.12 Selection of main and tender springs

- Helper (Figure 3.13) – these springs are used to prevent the main spring from becoming loose (slack) in the spring seat when the wheel is in full droop. Helper springs have a very low spring rate and do not affect the suspension characteristics; when under load, they fully compress. Typically up to 60mm can be taken up by a helper spring.

Figure 3.13 Helper spring – notice how its flat and thin coil construction allows it to easily compress flat

Spring specifications

Springs can be purchased in both metric and imperial form. The specification of a spring is normally engraved in the top or bottom surface of the spring or etched into the side of one of the coils.

A spring rate tester (Figure 3.14) is a good tool to ensure consistency across a set of springs, as manufacturing tolerances could mean a different spring rate on each corner of the car, resulting in handling irregularities and ride height/corner weight issues. High-end spring manufacturers, such as Eibach and Hyperco, are a safe bet for consistent springs – but it is advisable to regularly check and measure rates and open lengths, along with tagging each one to build a history of that spring.

Springs are generally rated by two different methods: newtons per millimetre (N/mm) or pounds per inch (lb/in) with a conversion factor from metric to imperial of × 5.71 (i.e. a 100 N/mm spring equates to 571 lb/in). The most common rating in the UK is lb/in.

A metric spring has a labelling system, for example 200-60-100, which translates into:

200 = Spring open length (mm)

60 = Internal diameter (mm)

100 = Spring rate (N/mm)

An imperial equivalent will use 7 × 2.25 × 325, which translates into:

7 = Spring open length (in)

2.25 = Internal diameter (in)

325 = Spring rate (lb/in)

Figure 3.15 shows three different types of spring set-up: a progressive main spring, linear main spring, and dual spring set-up utilising a linear main and tender spring. Here you can see how using different spring configurations could affect the car's characteristics.

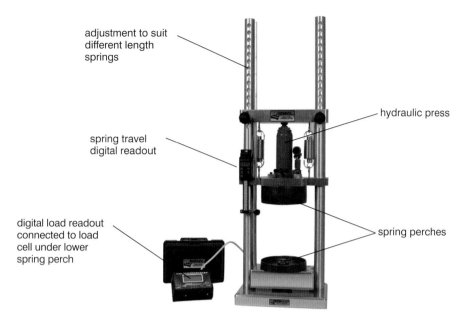

Figure 3.14 Longacre electronic coil spring tester

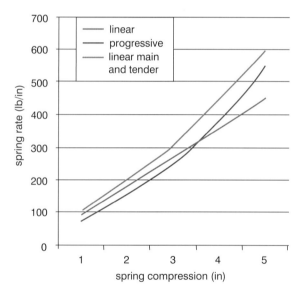

Figure 3.15 Comparison of different spring types

Preload

Preload is when a small amount of compression is applied to the spring before it is subjected to any additional compressive load from the car, hence the term 'preload' (before load). It describes the amount of compression a spring is under in its rest state. Minimal preload can be used to hold the coil spring in place at full extension if it means helper springs can be avoided.

However, preload can also be an effective tuning tool when adjustable push/pull rods are used to adjust ride height.

Preload is added by winding up the lower spring perch (Figure 3.16) to compress the spring. It can be measured by either counting the turns that the lower spring perch has taken to compress the spring or, more accurately, by measuring the difference between the spring's open length and its compressed length when preloaded.

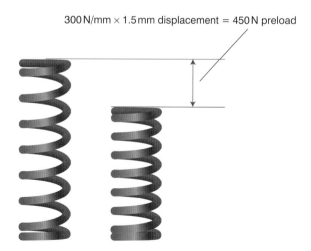

Figure 3.16 Spring without preload and with preload

For example, if you have a 100N/mm spring and compress it by 2mm, a preload of 200N/mm is added. The spring will not compress any further until a force that exceeds this preload is applied to the spring. This means that the spring will need approximately 20kg (newtons/gravity = kilograms) of vehicle weight to overcome the preload. Using imperial springs, the scenario is the same, but we just use different units. If we use a 300lb/in spring and compress it by 1/4in (0.25in), a preload of 75lb/in would be added.

So what does preload actually do?

As preload is initially applied it will have the simple effect of increasing the ride height. This can be compensated for with push/pull rod adjustment if required or available.

Preload limits droop travel as, when the load on the spring is less than the preload value set, the spring and damper assembly is effectively at full extension. This prevents the wheel from moving in rebound any further. If more than the corner weight of the car is applied in preload, the suspension will not move until sufficient aerodynamic load (downforce) is applied to 'break' the preload. This is sometimes referred to as running more than 100 per cent preload.

For example, if a car with a corner weight of 250kg had a preload added of 2500N/mm by compressing a 100N/mm spring by 25mm, this would equate to 250kg of force and, therefore, 100 per cent preload.

When the preload on the spring is not overcome, the suspension will not move. This has the effect of making the car very reactive, with an excellent change of direction. This can be particularly useful for low-speed corners, such as chicanes and hairpins.

Running more than 100 per cent preload in combination with bump rubbers allows the possibility of running three-spring stiffness. This would give excellent change of direction at very low speed (preload), good compliant mechanical grip in the mid-speed (spring rate composition) and increased stiffness at higher speed to support the aerodynamic load (spring rate and bump rubbers).

Torsion bar springs

Torsion bar springs are becoming more popular in motorsport. They utilise multi-link suspension systems, such as push/pull rod systems, and are often fitted in the pivot point of a rocking arm or bell crank (Figure 3.17). The sprung steel bar will have splines at either end; one end will be keyed into the chassis while the other will spline to the rocker or bell crank.

Some key points:

- In a torsion spring the elastic properties of a twisted bar are used to produce a linear spring rate, much like that of a coil spring.
- By altering the diameter of the bar, the corresponding linear spring rate is also changed.

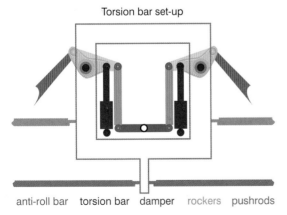
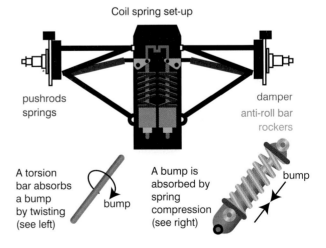

Figure 3.17 The rear suspension of a Formula 1 car showing the use of torsion bar springs and two dampers without coil springs around them on the left, and a coil spring set-up on the right

- A larger diameter bar gives a stiffer linear spring rate.
- The main advantages of torsion bar suspension are durability, easily adjustable ride height, and taking up less of the vehicle's interior volume compared to coil springs.
- A disadvantage is that torsion bars, unlike coil springs, usually cannot provide a progressive spring rate.

Selecting spring rates

It is vital to get this right and can save hours of wasted testing at the race track. The first stage is to select the ride frequency (undamped natural frequency of the body in ride) of the car's front and rear – the higher the frequency, the stiffer the ride rate.

There is often a difference between front and rear frequencies and there are different theories for this:

- Theory 1 – the rear must be higher as it has to catch up with the front when tackling the same bump in a road.
- Theory 2 – the front is higher, which allows faster response on turn in, less ride height variation (keeping the aerodynamics and geometry stable) and better rear-wheel traction for corner exit. A 10–20 per cent difference is often found between the front and rear ride frequencies. Figure 3.18 shows an example of front and rear ride frequency.

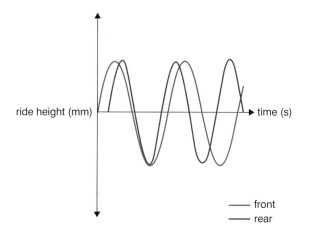

Figure 3.18 Higher rear ride frequency

A rough guide to picking the frequencies:

- 0.5–1.5 Hz for passenger cars
- 1.5–2.0 Hz for saloon race cars and moderate downforce formula cars
- 3.0–5.0+ Hz for high downforce race cars

Once this is done, the following calculation can be used to select spring rates.

$$Ks = 4\pi^2 \times Rf^2 \times Sm \times MR^2$$

Where:

Ks = Spring rate (N/m)

Rf = Ride frequency (Hz)

Sm = Sprung mass (kg)

MR = Motion ratio (wheel/spring travel)

For example, with a sprung mass of 150 kg, ride frequency of 3.5 Hz and a motion ratio of 1:1, we can calculate:

$$Ks = 4\pi^2 \times 3.5^2 \times 150 \times 1^2 = 72542 \text{ N/m}$$
$$= 73 \text{ N/mm}$$

or approximately 414 lb/in

Motion ratio

The most vital thing to know is wheel rate. Wheel rate is effectively what the driver feels and the tyres deal with while driving. It takes the geometry of the suspension into the equation (motion ratio) and makes it easy to compare different cars, like for like. The car's wheel rate characteristics are determined by the motion ratio.

Motion ratio is very simple to find. All you have to do is jack the wheel up one inch and measure how far the lower spring perch has moved relative to the upper spring perch and do this until the spring/damper runs out of useable travel. Below are the calculations required to find motion ratio and wheel rate.

$$\text{Motion ratio} = \frac{\text{Wheel travel}}{\text{Spring travel}}$$

$$\text{Wheel rate} = \frac{\text{Spring rate}}{(\text{Ratio})^2}$$

Table 3.3 Increasing wheel rate due to spring axis geometry

Wheel travel (in)	0–1	1–2	2–3
Spring travel (in)	0.60	0.65	0.70
Motion ratio	1.67:1	1.54:1	1.43:1
(Motion ratio)^2	2.79	2.37	2.04
Wheel rate for 412 lb/in spring (lb/in)	148	174	202

Table 3.4 Decreasing wheel rate due to spring axis geometry

Wheel travel (in)	0–1	1–2	2–3
Spring travel (in)	0.65	0.60	0.55
Motion ratio	3.54:1	3.66:1	3.82:1
(Motion ratio)^2	2.37	2.78	3.31
Wheel rate for 340 lb/in spring (lb/in)	148	126	106

A graph can be plotted to discover whether the suspension is rising rate, falling rate or linear. Tables 3.3 and 3.4 show the difference between a rising and falling rate system.

Figure 3.19 shows two different motion ratio measurements.

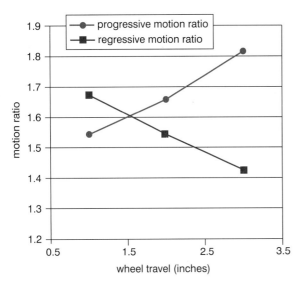

Figure 3.19 Motion ratio change due to suspension geometry

Due to the inclination of the spring axis and associated linkages, the motion ratio is unlikely to remain constant as the spring compresses. It can increase, decrease or stay linear. Ideally, the wheel rate increases slightly as the spring compresses, or at least remains linear. Progressive wheel rates can also be achieved using progressive springs or bump rubbers.

3.2.3 Bump rubbers

Bump rubbers provide another source of adjustability and are found fitted on the damper shaft. The bump rubber provides a spring rate that is in parallel and, therefore, the effective spring rate of the assembly will be the actual spring rate plus the spring rate of the bump rubber (which is often progressive itself).

They are most commonly used to prevent the damper bottoming out but can be used in a sophisticated manner to control aerodynamic grip when the downforce of the car is taking up vital damper travel.

Bump rubbers come in a range of specifications (lengths, sizes and materials) that can determine their initial stiffness, the point at which the spring rate begins to rise, and the rising rate itself.

3.2.4 Dampers

These are significant tuning tools and are often the first point of adjustment for most race cars once the base set-up is found (including geometry, corner weights, etc.).

Technical manual 2812, 2816, 2817

BUMPSTOP INFORMATION

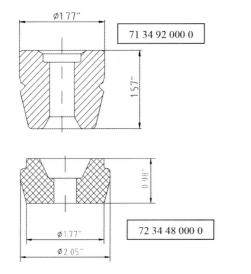
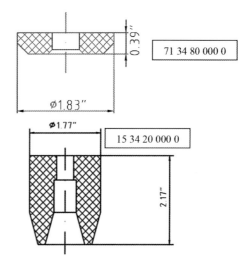
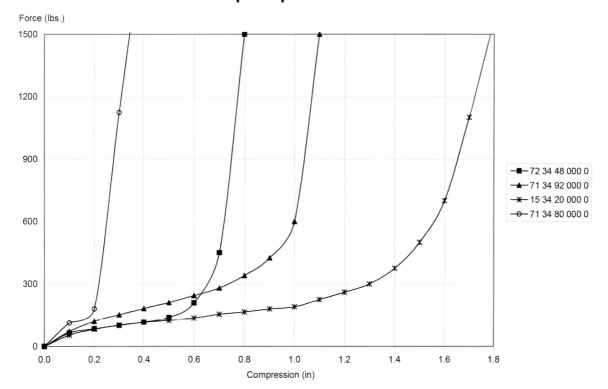

Figure 3.20 Koni bump stop technical data

The dampers control the speed of the springs' movement and, therefore, control how the suspension moves. The damper acts as a grip adjustment tool and can also have an impact on tyre degradation. It converts the kinetic energy from the spring into heat energy, which is absorbed by the oil inside the damper.

Dampers operate in quite a simple manner. One end of the damper is connected to the chassis, while the other end is connected to the moving suspension, directly or through a linkage. As the damper moves up and down, a piston within the oil chamber moves up and down and the oil is forced through a series of holes. The size of these holes determines how hard or soft the damper is and how slowly or fast the system can move. The valving in the piston has a series of calibrated shims that deflect at certain forces, and it is this deflection that allows the oil to pass through.

Dampers come in a variety of options, such as open and closed lengths, different material construction, mounting type, and eyelet and shaft sizes.

You can also choose the range of adjustment you would like, which also determines the cost of the unit:

- One-way adjustable (bump and rebound together at a fixed ratio)
- Two-way adjustable (bump and rebound separately)
- Three-way adjustable (high- and low-speed bump and rebound)
- Four-way adjustable (high- and low-speed bump and high- and low-speed rebound)

Low-speed and normal bump and rebound controls the sprung mass and so tunes roll and pitch, while high-speed bump and rebound controls the unsprung mass, tuning how the car reacts to bumps and kerbs. This allows you to tune the car, not only to the requirements of the chassis balance, but also to the quality of the surface being raced on.

The valving inside a damper can be altered to tune the switchover period between the high- and low-speed characteristics of the car.

Dampers built with a pressurised nitrogen cartridge have a better ability to avoid cavitation of the oil. Cavitation is when air bubbles form in the oil when the oil is accelerated quickly and the gas pressure chamber (if present) does not equalise the pressure differential on either side of the piston. This reduces the damping forces that can be produced and, therefore, makes the handling of the car inconsistent.

Gas-filled monotube dampers are the most common high-end motorsport dampers, along with the external reservoir system. Through-rod dampers are also considered to be very good, but have not been seen to be worth the extra cost over the previously mentioned dampers.

Twin tube dampers are often the cheapest but, when worked hard, the oil inside can often foam and give an inconsistent damper.

A monotube damper has some benefits due to the volume of oil that is exposed to the outer wall of the damper and so can often cool better. A gas-filled monotube damper has a high-pressure chamber of nitrogen (gas reservoir) at the bottom, which makes it difficult for the foam/bubbles in the oil to form and so produces a more consistent damper.

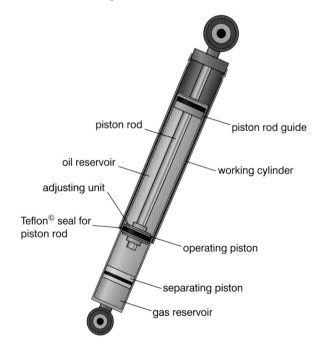

Figure 3.21 Monotube gas-filled damper layout

External reservoir dampers use an additional reservoir so that more oil can be carried, which makes the cooling capacity greater as well as reducing unsprung weight because the reservoir can be chassis-mounted. As the oil flows to and from the shock body, adjustability can be precisely controlled.

Damper quality is important. The cost is determined by the quality and the adjustability built into the damper, and the prices set by the manufacturers vary greatly. The quality and consistency of the shims and internal machining is what makes dampers consistently repeatable. Performance is also gained through low levels of friction, which is determined through seal technology and machining tolerances.

A damper dynamometer can test damping forces and also match dampers to ensure that pairs are the same as one another. A 'spud plot' compares load against displacement to show how the damper reacts in bump and rebound at every different adjustment. Force against velocity will represent the curve of the dyno in terms of how it reacts at different linear speeds.

Figure 3.22 Damper with external reservoir

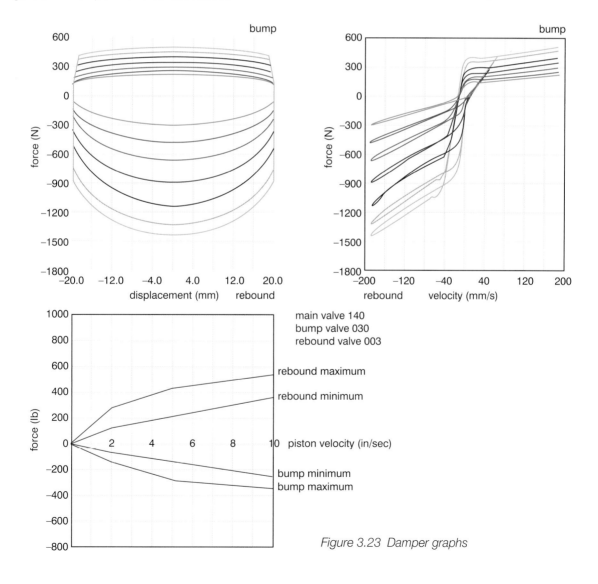

Figure 3.23 Damper graphs

3.2.5 Anti-roll bars

Anti-roll bars (ARBs), commonly used in race cars, perform two functions. Firstly, they control and limit body roll and, secondly, they can be used to tune the car's handling balance.

They can be adjusted by altering their leverage points (outer mounting points), by changing the bar for a different thickness version, or if a blade ARB is in use, rotating it to change its strength. They do not stop weight transfer, they merely reduce body roll.

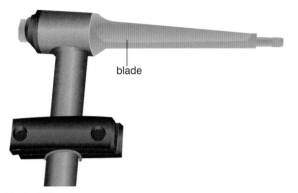

Blade ARB

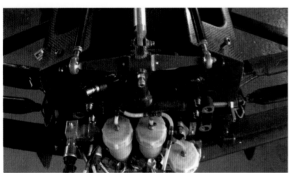

T-bar

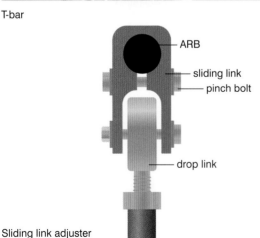

Sliding link adjuster

Figure 3.24 Various ARB layouts

In terms of handling balance, it is a general rule that increasing the stiffness at one axle will increase the grip at the other end of the car.

One downside to ARBs is that, as they connect one wheel to the opposing side, they can reduce the independence of the axle, which can make tackling bumps and kerbs tricky.

3.2.6 Geometry

Geometry is another key factor of chassis set-up.

Camber

Measured in degrees, this is the angle of the wheel from the front view in comparison to the vertical. Negative camber is used in motorsport to provide a larger contact patch when cornering. Static camber and dynamic camber change through suspension travel. It is used to provide optimum lateral grip when cornering. However, running high amounts of camber can reduce braking and traction ability, while it also has the potential to overheat and overload the inside shoulder of the tyre. Large amounts of camber also make the car hard to drive on cold tyres.

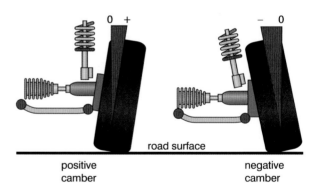

Figure 3.25 Camber

Toe

Measured in millimetres, this is the direction in which the wheels are pointing when viewed from above. Toe has the ability to tune the car's stability and responsiveness. On a rear-wheel drive car, it is common to run a small amount of toe out on the front to provide a good initial

Suspension

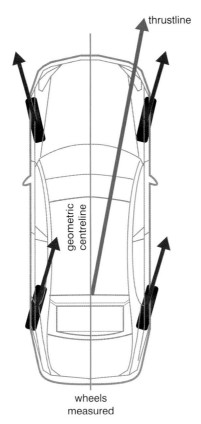
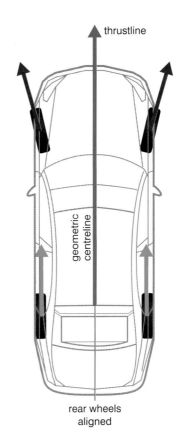

Figure 3.26 Toe and alignment

turn in, while a small amount of toe in at the rear creates a stable rear end. Running toe also provides a small amount of scrub and, therefore, helps to generate tyre temperature. Alignment of the rear wheels is also important to remove any thrust angles, which is when the rear wheels do not track with the fronts and allow crabbing to occur.

Caster

Measured in degrees, this is the angle of the steering axis compared to a vertical line through the centre of the wheel when viewed from the side. Positive caster is used to provide an increase in negative camber when steering. Too much can make the car difficult to drive, because steering can become heavy and too reactive as it tries to return to its central position.

Other suspension geometry that is usually designed into the car and cannot be easily changed individually is covered on the following pages.

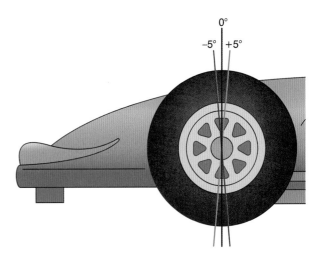

Figure 3.27 Caster angle

Roll centre

The roll centre is the point that the sprung mass rotates around. This is based on the geometric angles of the wishbones and can be visualised by theoretically extending the wishbone lines until they meet at the instantaneous centre (IC) point. From there, a line is drawn back across the car to the centre of the tyre contact patch. The roll centre is where this line crosses the centre of the car. A high roll centre gives less body roll but more jacking, while a low roll centre gives more body roll and less jacking. The body roll assumed with a low roll centre can be controlled with an ARB. The skill in designing a suspension system includes trying to ensure that the roll centre at both axles does not migrate wildly while under a dynamic state.

Roll axis

The roll axis is a line linking the front roll centre to the rear roll centre. In most cases the front roll centre is often lower than the rear roll centre. A roll axis needs to remain at a consistent inclination to provide a stable and consistently handling car. By altering the heights of the roll axis, you will affect the roll moment distribution between the front and rear axles.

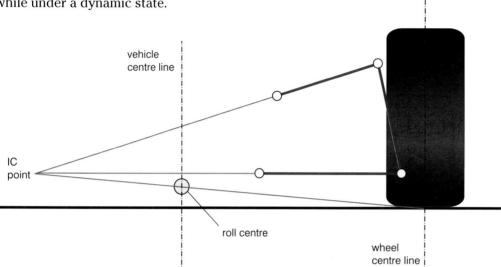

Figure 3.28 Roll centre location

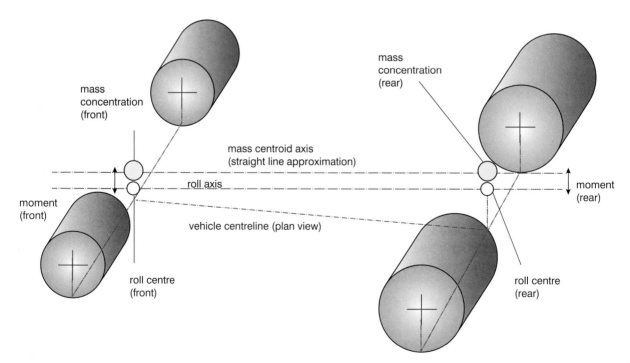

Figure 3.29 Roll axis, moment and centre of gravity (CG)

Roll moment

This is the inertia of the sprung mass of the car during roll. There is a front and rear roll moment, and this is determined by the distance between the CG and roll centre at that axle. By situating these two points close together, body roll will be reduced considerably.

Centre of gravity (CG)

This is one of the most important parts of the car's design. The CG is an imaginary point in which all forces act through on the car and is effectively the point at which the car would remain balanced if pivoted on. The lower the CG the better, as this will provide less weight transfer in all dynamic states and allow each tyre to deal with a more even amount of loading.

Weight transfer

The ideal scenario is for the car's axles to have an evenly loaded tyre at either side so that each tyre can be worked to its maximum potential. However, with weight transfer acting through the CG, this is not possible. Weight transfer occurs laterally when cornering, longitudinally during accelerating and braking, and diagonally when combining the two together.

With weight transfer, we can conclude the following:

- The more g-force a car can produce, the quicker it will cover a lap. Cornering g-force can be dictated by downforce, tyre contact patch and compound, suspension geometry and track width. Accelerating g-force can be determined by engine performance, tyre contact patch and compound, and gear ratios. The surface of the track also makes a difference. Braking g-force can be determined by braking power (disc size, etc.), tyre contact patch and compound. You should be starting to understand the importance of the tyre by now.
- The heavier the car, the more weight will be transferred from the inside to the outside, reducing the total grip of the car. The opposite occurs for a lighter car.
- The higher the CG, the more weight is transferred; equally, the lower the CG, the less weight is transferred.
- The wider the track, the less weight transfer; however, the drawbacks of a wide track are a larger frontal area and, potentially, a heavier, less nimble car.
- The wheelbase of the car is often fixed in its design, but can provide a change in static weight distribution and alter weight transfer. However, again, this adds to the mass of the car and potentially reduces the responsiveness of the car's handling.

3.2.7 The 'antis'

Anti-pitch geometry controls the amount of pitch movement of the car by reducing the amount of load going through the suspension springs, and placing a proportion of the load transfer through the suspension arms instead. There are several types of anti-pitch geometry:

- Anti-dive – this restricts the nose of the car diving under heavy braking.
- Anti-squat – this restricts the rear of the car dropping down under hard acceleration.
- Anti-lift – this can also be used at either end to stop the front moving into droop in acceleration, while at the rear it stops it going into droop during braking.

The benefit of the 'antis' is that the change in geometry through suspension movement and ride height can be reduced through using it. However, too much 'anti' and the car will lose its compliance and be difficult to drive.

Antis are measured as a percentage. To determine the percentage of rear suspension braking anti-squat, as shown in the Figure 3.29, it is first necessary to determine the tangent of the angle between a line drawn, in side view, through the rear tyre patch and the rear suspension IC, and the horizontal (expressed as h). Then, divide this tangent by the ratio of the CG height to the wheelbase (expressed as l). Finally, multiply by 100.

As an example, a value of 50 per cent would mean that half of the weight transfer to the rear wheels, during acceleration, is being transmitted through the rear suspension linkage and half is being transmitted through the rear suspension springs.

Front suspension anti-dive is calculated in a similar manner and with the same relationship between percentage and weight transfer.

$$\text{Anti \%} = \frac{\tan \theta}{(h \div l)} \times 100$$

Example:

$$\text{Anti \%} = \frac{\tan 11.3}{(0.5 \div 2.5)} \times 100$$

$$\text{Anti \%} = 100\%$$

3.2.8 Wishbone layout

The double wishbone system is the most commonly used suspension for race cars. It provides freedom of design and adjustability. Wishbone layouts can vary from parallel to non-parallel and equal length to unequal length. The total length of the wishbone can also provide different operating characteristics. The choices determine camber change, track change and wheel control during dynamic travel. Below is a summary of these layouts.

Equal and parallel

The roll centre is on the ground, so there is no camber change in bump/rebound. There is a change in track width during bump/rebound. The outside wheel changes towards positive camber during chassis roll by the same angle as the chassis roll. Longer wishbones help to reduce the camber and track change but do not completely remove it.

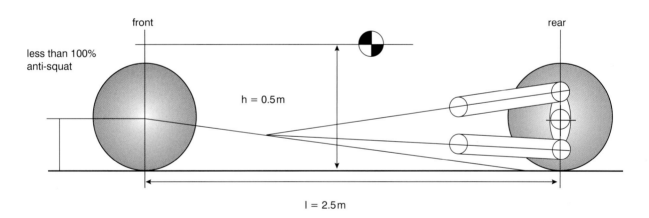

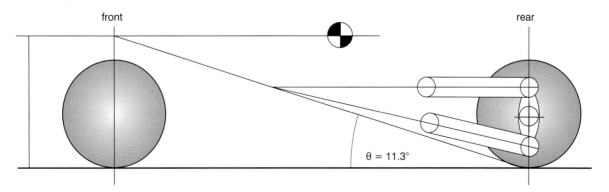

Figure 3.30 Anti-geometry

Unequal and parallel

The upper wishbone is shorter in length than the lower one. This creates a different wheel path, which results in negative camber increasing in bump and a change to positive or negative in droop based on layout. There is minimal track change. During roll, the outside wheel will lose a small amount of negative camber, while the inside wheel tends to gain more negative camber. Roll centre location is stable, which provides a good roll moment. The roll centre is still assumed to be ground level.

Unequal and non-parallel

The roll centre can be placed anywhere when static (although the skill is keeping it there or at least close during dynamic suspension movement). Inclining the wishbones in towards the chassis provides better control of camber change in roll. There is minimal track change and more freedom of camber curves in vertical movement and roll.

Swing axle lengths are determined by the distance from the roll centre IC point, as discussed earlier, back to the tyre. This dimension is another factor in wishbone design and its length can characterise roll centre location/migration and camber/track change in vertical/roll conditions.

Kinematic software allows you to easily analyse suspension design; a simple set of measurements to apply some coordinates to the software package allows you to dynamically analyse the performance related to various suspension designs in order to select the optimum design for your application.

The software will also help you to determine a suspension system to have a linear, rising or falling rate based on its geometry. This means that the motion ratio (see section on springs) changes during suspension travel, and this can then effectively change the wheel rate of the car during travel.

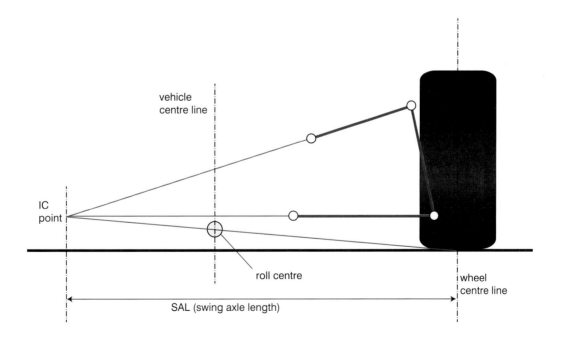

Figure 3.31 Swing axle length (SAL) view

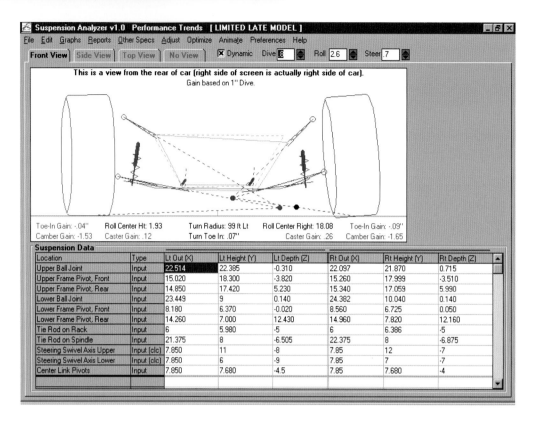

Figure 3.32 Suspension analysis software

3.2.9 Spring/damper activation linkage systems

This is a brief overview of the main configurations used to activate the spring/damper assembly in modern race cars.

Direct to wishbone and rocking arms were common until packaging, aerodynamics, preload adjustment and motion ratios became more important. The majority of race cars from the 80s onwards began to use the push/pull rod system.

Push/pull rod system

The pull rod system operates a rod linking the bottom wishbone to a rocker/bell crank and then to the damper. This allows the damper to be placed in any location (normally just above the driver's legs or on top of the gearbox), which helps with packaging a compact race car, accessibility and CG. Aerodynamic drag is reduced as the outboard spring/damper assembly is replaced for an aerodynamic profiled rod. Control over motion ratio is pretty much infinite, while ride height and corner weight adjustment is a lot quicker and easier to achieve. The pushrod can be fitted with left- and right-hand rose joints, so that when the lock nuts are slackened, the rod can simply be rotated one way or the other to set ride heights/corner weights. Pickup points for the rods usually come from the outermost point of the wishbone or from the upright itself. Whereas the pushrod pushes from the bottom wishbone, the pull rod pulls from the top wishbone and often has the spring/damper unit situated nearer the floor of the car. The pull rod can give benefits of a slightly lower CG but access can be difficult; and their use in Formula 1 can be limited due to the high nose not allowing for a low slung damper at the front, as geometry conditions can be limited.

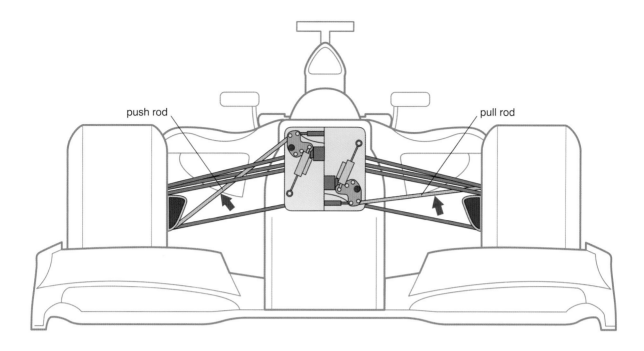

Figure 3.33 Push/pull rod systems

With the common system of a push/pull rod operating a pair of rockers and spring/dampers, ARBs are often then operated from the rocker to a T-bar mounted in the chassis or a conventional n-shaped bar that links either side. A sliding link, blade bar or multiple holes allow for a different stiffness to be used.

Third spring set-up

An anti-heave system can also be used to control aerodynamic downforce affecting the ride height of the car. This is normally via a third spring/damper set-up. The amount of suspension travel will determine whether a damper is needed instead of just a spring (coil or Belleville).

Monoshock system

The monoshock system is commonly found on the front of Formula-style cars. This utilises one damper/spring unit.

This unique set-up uses a push/pull rod system, and both rods are mounted on a common rocker, itself mounted on a base plate on the chassis and able to move laterally and rotate around a lateral axis. Either side of the common rocker is a Belleville spring stack. When the car operates in roll, the common rocker slides sideways and the springs control the roll. Note that the roll movement is undamped, suiting this set-up only to cars with minimal roll (e.g. Formula 3, etc.). In pitch movements, the rocker rotates and actuates the sole spring/damper unit, which is mounted at one end to the common rocker and the chassis at the other.

The key benefit of this system is separate pitch and roll rates that are infinitely tuneable. The Belleville stack can be configured in a variety of ways based on spring type and orientation and the preload added to them. This system is not used on the rear of race cars due to the traction issues that could occur from an undamped and stiff rear set-up (although the third spring set-up acts in a similar way), but the extra springs add more cost and complexity to the system. Although the system seems to have a drawback in that it has no damping in roll, we would assume that it also would lack grip on a less than smooth circuit, but this seems to not be the case with its common use in sprint and hill climbs. When the car moves into roll, the outer wheel is forced upwards, and pushes the inner wheel into the ground to provide great turn in grip.

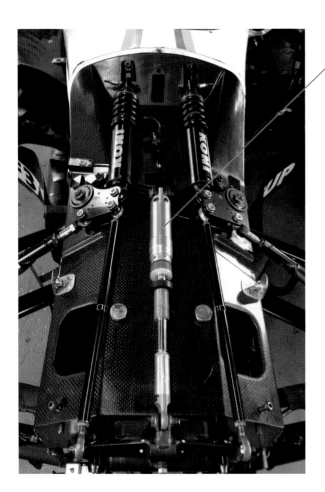

Figure 3.34 Three damper set-up

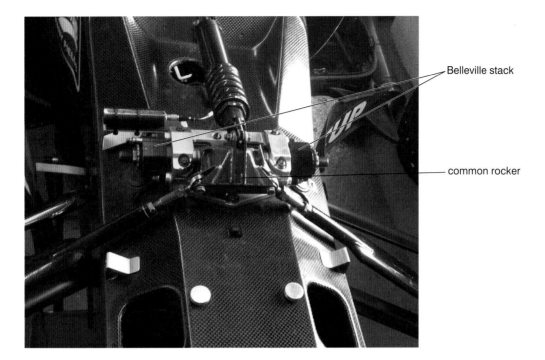

Figure 3.35 Monoshock system

3.3 Steering

The steering system is very limited in its 'tuning' capacity if we assume that 'toe' is a suspension-based adjustment.

The physical part of the system obviously starts at the steering wheel itself. The size of the wheel determines the torque that the driver can feed into the steering. A smaller steering wheel will require more effort to turn but will give more feedback, whereas a larger wheel will be easier to turn but would potentially give less feedback. A flat bottomed, D-shaped wheel will give more space for the driver's legs but size can often be dictated by the cockpit space and driver preference.

A quick-release (QR) boss allows the steering wheel to be easily removed to help the driver get in and out of a cramped cockpit. The splined set-up, with a master spline, is the best choice over a hex-shaped boss as it allows the wheel to return to the same position each time.

The steering wheel needs to be positioned so that it reduces stress on the driver's wrists and enables comfortable arm reach. Outstretched arms will fatigue and make control difficult, while being too close will limit the driver's movement.

A steering rack will generally need to have a quick ratio rack in order for the car to be responsive and to allow the driver to apply a large steering angle over a relatively small steering wheel angle: the quicker the rack, the heavier the steering. Lock stops can be attached to the rack gear where it exits the casing. These prevent the steering from overthrowing and fouling (hitting) the bodywork or suspension components. We can calculate either steering rack ratio or overall steering ratio:

- Steering rack ratio – turn the steering wheel one turn and then measure how far the steering rack has moved. For example, 2 inch rack travel to one steering wheel rotation = 2:1.
- Overall steering ratio – turn the steering wheel one turn and use the turn plate to measure the angle of the inside wheel (Road wheel angle ÷ 360). For example, steering wheel turned 360°, road wheel turns 36°, so 36 ÷ 360 = 10:1 (10° steering wheel movement to 1° road wheel movement).

Figure 3.36 Steering wheel and quick-release (QR) boss

3.3.1 Bump steer

This is a scenario whereby the road wheels steer themselves without any input from the steering wheel. This occurs when bumps in circuits and kerbs on the circuit cause the steering and suspension to react undesirably based on the angles of the linkages of the two systems.

Most race cars are designed so that there is a minimal amount of bump steer. However, when maintaining or rebuilding any steering and suspension assembly, you must ensure that the correct shims and spacers are placed back in the same position in order to avoid unwanted bump steer. Using the correct components from the right make and model of car (unless it is a proven upgrade package) is essential. Bump steer should be designed into the car and cannot always be easily adjusted once the car has been built. There is specific set-up equipment available to measure bump steer. Suspension, steering mounting and pivot points, if moved, must take into consideration the effects of bump steer.

The front-end of the car must be designed correctly in order to achieve no bump steer. The track rod has to travel in the same arc as the suspension wishbones during both bump and rebound travel. It can be done by simply matching the lengths and arcs of both systems to prevent any unwanted bump steer of the front tyres. Bump steer can be fine-tuned by altering rack height or by adjusting the positional height of the track rod end – some cars often have adjustment in the form of shims either side of the track rod end. A bump steer gauge can be used to check and adjust it. While zero bump is desirable, if this cannot be achieved, it is preferable to have a very small amount of toe out in bump, which will give some response and good turn under braking and starting a turn at a corner.

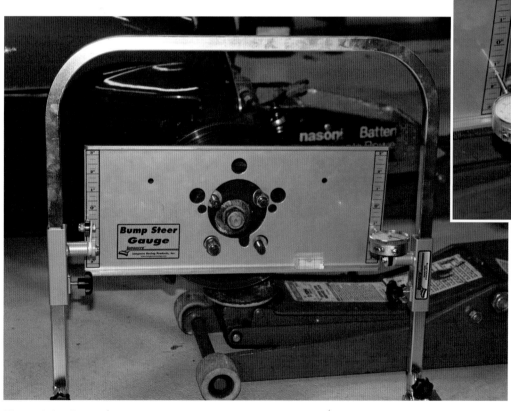

Figure 3.37 Bump steer gauge and adjustment

3.3.2 Ackerman steering principle

There are three types of Ackerman (see Figure 3.38), all of which vary in popularity:

- Ackerman – all four wheels travel along an arc that has the same common instantaneous centre. As a result, it is assumed that both front tyres are travelling at the correct geometric radius to avoid any form of scrub.
- Pro-Ackerman – the front wheels have a larger combined steer angle than Ackerman steering, and the inner wheel turns at a tighter radius than the outer. Pro-Ackerman is useful on tracks with multiple low-speed tight corners.
- Anti-Ackerman – the front wheels have a smaller combined steer angle than Ackerman steering, and the outer wheel turns at a greater radius than the inside. A race tyre's maximum lateral grip for a given vertical load generally occurs at higher slip angles as the vertical load increases. This means that for both the inside (less vertical load) and outside (more vertical load) tyres to be at their maximum available grip, the outside tyre needs to be at a higher slip angle than the inside tyre (see Figure 3.39). Anti-Ackerman achieves this by causing the outside tyre to be at a larger steer angle than the inside tyre.

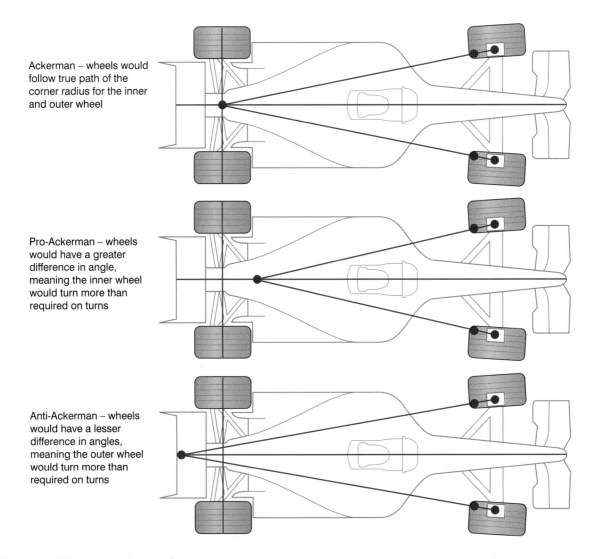

Figure 3.38 The three Ackerman layouts

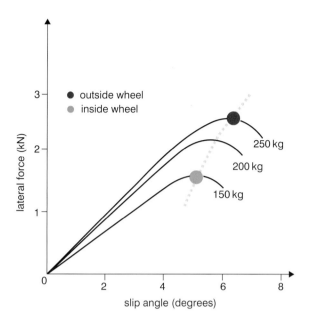

Figure 3.39 Lateral load vs. slip angle

3.4 Brakes

The braking system is undoubtedly important in any race car. Time and places are gained in the braking zones so, although we spend less time braking than doing anything else, it is still vital. The brakes must not only provide stopping power, but must also provide feel, consistency and modulation. It is for this reason that there is a large amount of research and development into braking systems and technology is rapidly advancing with the development of new materials and designs.

Again, we must refer back to the all-important tyre, without which the brakes would be of no use.

3.4.1 Braking basics

A well-designed braking system will allow the driver to modulate the braking effort through force (pressure) on the brake pedal and not by displacement of the pedal. The brake pedal must be very firm with limited travel – this requires stiffness through the whole system including all mounting points and ensuring that pedal ratios, master cylinder (MC) sizes and calliper piston sizes are correct.

Since it requires the same energy to decelerate from a given speed as it does to accelerate to that speed, a huge amount of energy is required to slow or stop a speeding vehicle. The brakes achieve this by converting the kinetic energy to thermal energy, which is then dispersed into the airstream.

3.4.2 Pedal box and MCs

The pedal box generally houses the brake pedal, clutch pedal and throttle and provides

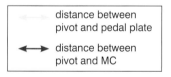

Figure 3.40 Pedal box assembly

a location for the MC(s) to be mounted. The brake pedal is the only point of contact between the braking system and the driver. Therefore, it must be designed to offer the most comfortable operating position without compromising rigidity in order to translate as much 'feel' as is possible back through the system. The pedal is mounted on a pivot and the MC(s) are located on the pedal to offer a mechanical advantage to the driver when pressure is applied to the pedal. Figure 3.41 illustrates the mechanical advantage pedal arrangement (note how the distance between the pedal plate and the pivot is longer than the distance from the MC to the pivot).

A larger pedal ratio will provide a greater pressure into the brake system for the same applied effort on to the brake pedal but longer pedal travel. This mechanical advantage can be altered and tailored to suit the vehicle, the brake system or the driver's preference, with relative ease.

Adjustability of the pedals and their positions is paramount in any racing car and must be designed into the system in the early stages in order to optimise the effectiveness. No two drivers will have the same size feet or the same preferences, nor will it be possible to perfectly position the pedals without first spending some time on the track, as race cars tend to feel very different when stationary compared to being driven at speed around a circuit. This adjustability can be achieved with threaded foot plate mountings or multiple mounting holes for the pedal box itself.

For motorsport applications there are a multitude of off-the-shelf pedal boxes; the choice can depend on the layout of the chassis or the requirements of the vehicle. In almost all cases they offer the use of dual MCs. Figure 3.41 shows a twin MC set-up incorporating a brake bias bar.

The dual MC separates the hydraulic line system from the front and rear of the vehicle. This provides a number of advantages. Firstly, in the event of a system failure at either the front or rear of the car (e.g. a split or burst brake line or calliper seal failure) the vehicle will still have a functioning braking system at the other end, permitting a safe return to the pits. Secondly, the pressure that is applied to the brake pedal can be shared between the two MCs in an unequal portion by the use of a bias bar. The

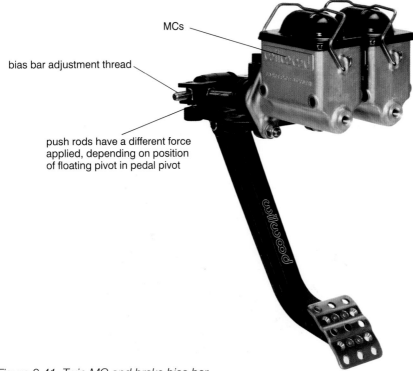

Figure 3.41 Twin MC and brake bias bar

brake bias bar allows almost infinite adjustment of the bias between the front and rear brakes and the system can be set up in such a way as to allow the driver to alter the bias from the seat.

Using a smaller MC than standard will give the following characteristics:

- Force required to depress the brake pedal is less (pedal feels softer).
- Pedal travel increases to achieve the same brake pressure.
- Some 'feel' can be lost in braking areas where a more subtle technique is required.

Using a larger MC than standard will give the following characteristics:

- Force required to depress the brake pedal is more (pedal feels harder).
- Pedal travel decreases to achieve the same brake pressure.
- Some 'feel' can be increased in braking areas.

If all four tyres have equal braking power then, under straight line braking on a smooth road, hard braking would cause the rear wheels to lock due to forward weight transfer.

When braking there will always be a load transfer from the rear of the car to the front. This is due to, and affected by, a number of factors, including wheelbase and CG height. Most mid-engined cars will have a rearward weight distribution, which helps to even the loadings on each corner of the car when braking, but weight will still ultimately shift forward.

This weight transfer when braking alters the load experienced between the tyres and the road, and since a tyre's tractive capability is a direct function of vertical load, this increase in load at the front and decrease in load at the rear (compared to steady state) will invariably mean a change in the capacity for braking at each end of the car. The rear, being reduced in load, will have less braking capacity than the front.

Figure 3.42 Weight transfer forces the front of the car to dive under braking (pitch movement)

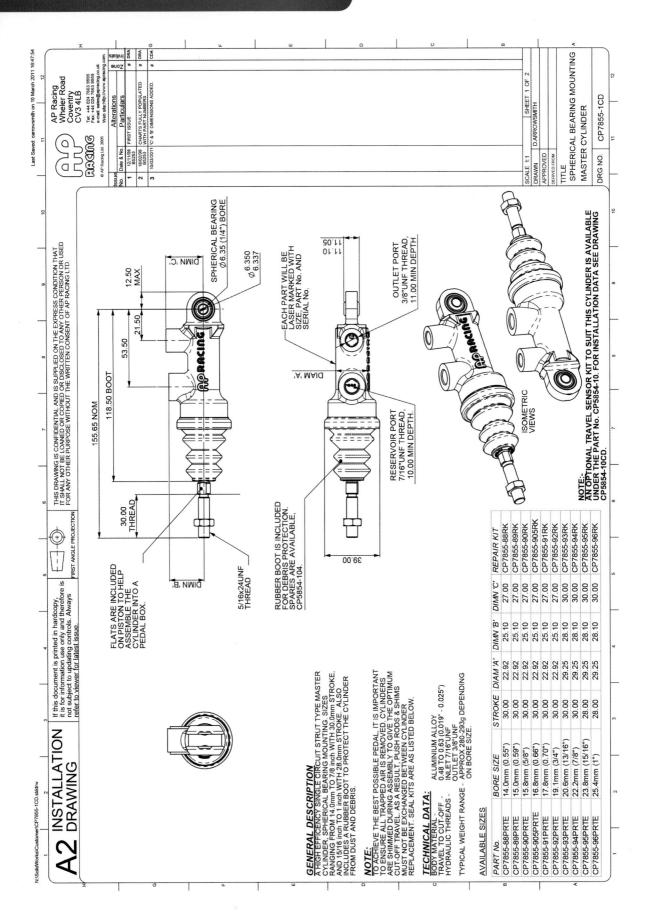

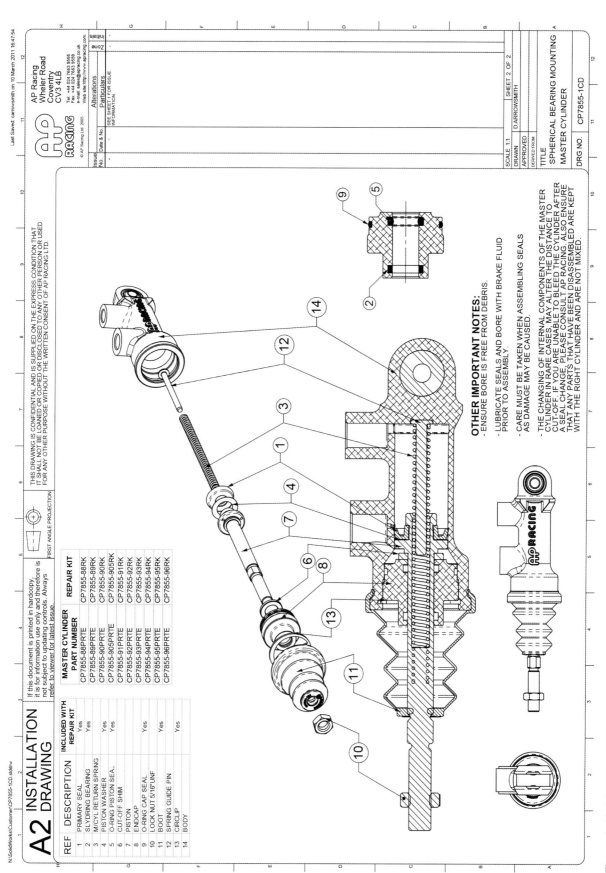

Figure 3.43 Master cylinder

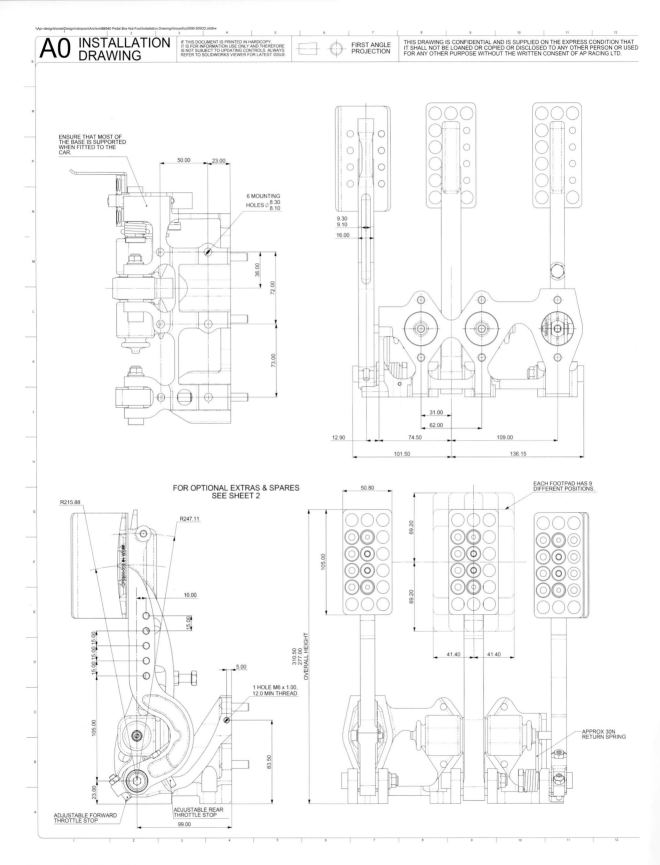

Figure 3.44 Pedal box

Brakes

GENERAL DESCRIPTION
THIS PEDAL BOX IS A GENERIC RACING PEDAL BOX DESIGN. DESIGNED FOR COMFORT AND CONTROL. THIS UPDATED PEDAL BOX HAS A NEW OPTIMISED MACHINED FROM BILLET BASEPLATE AND PEDALS. THE THROTTLE HAS BEEN UPDATED TO ADD ADDITIONAL FEATURES TO AID CONNECTION TO BELL CRANKS AND CABLES.

FEATURES

- OPTIMISED, LIGHTWEIGHT, MACHINED FROM BILLET BASE PLATE.
- OPTIMISED, LIGHTWEIGHT CLUTCH AND BRAKE PEDAL, WITH IMPROVED TWIST DISPLACEMENT.
- NEW FORGED THROTTLE PEDAL WITH ADDITIONAL FEATURES.
- ADJUSTABLE RATIO PEDALS.
- ALL THREADS ARE METRIC, EXCEPT THOSE MENTIONED BELOW.
- Ø10 BALANCE BAR
- ADJUSTABLE STOPS ON CLUTCH AND THROTTLE PEDALS.
- PEDAL BOX WEIGHT IS 2.50 Kg
  (CP2623 MASTER CYLINDERS APPROX 0.31Kg EACH)
- CABLE ADJUSTER INCLUDED.
- ALL PEDALS HAVE BALL BEARING PIVOTS.

PART NUMBER

CP5500-505MET - METRIC, PEDAL BOX WITH BRAKE, CLUTCH AND THROTTLE

CP5500-505UNF - IMPERIAL, PEDAL BOX WITH BRAKE, CLUTCH AND THROTTLE
THE ONLY THREADS THAT ARE IMPERIAL ARE THE 3 CLEVIS'S THAT ATTACH TO THE MASTER CYLINDER PUSHRODS.

BORE SIZE	BRAKE PEDAL	CLUTCH PEDAL
Ø14.0mm (0.551")	CP2623-88PRM088	CP2623-88PRM088
Ø15.0mm (0.591")	CP2623-89PRM088	CP2623-89PRM088
Ø15.9mm (0.625") 5/8"	CP2623-90PRM088	CP2623-90PRM088
Ø16.8mm (0.661")	CP2623-905PRM088	CP2623-905PRM088
Ø17.8mm (0.700")	CP2623-91PRM088	CP2623-91PRM088
Ø19.1mm (0.750") 3/4"	CP2623-92PRM088	CP2623-92PRM088
Ø20.6mm (0.842") 13/16"	CP2623-93PRM088	CP2623-93PRM088
Ø22.2mm (0.875") 7/8"	CP2623-94PRM088	CP2623-94PRM088
Ø23.8mm (0.937") 15/16"	CP2623-95PRM088	CP2623-95PRM088
Ø25.4mm (1.000")	CP2623-96PRM088	CP2623-96PRM088

THIS TABLE SHOWS THE RECOMMENDED PUSHROD LENGTHS FOR CYLINDERS IN THIS BOX.

IF YOU ARE ORDERING THE CP5500-505UNF, THEN MAKE SURE YOU ORDER YOUR CYLINDERS WITH THE 'PRT' EXTENSION AND NOT THE 'PRM', eg CP2623-91PRT088

CP2623 CYLINDERS HAVE IMPERIAL THREADS ON THE INLET AND OUTLET, FOR METRIC THREADS ORDER CP5623 VERSIONS INSTEAD.

CYLINDERS WITH LONGER PUSHRODS CAN BE CUT DOWN.

CP2905-18 CABLE ADJUSTER,
THIS IS INCLUDED WITH THIS PEDAL BOX

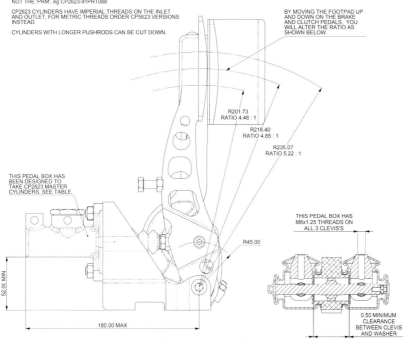

TITLE: GENERIC FLOOR MTG PEDAL BOX
DRG NO. CP5500-505CD
SCALE 1:1 **SHEET** 1 OF 3

Chassis

SPARES LIST

ITEM NO.	PART NUMBER	DESCRIPTION	Default QTY.
1	CP2494-1305	WASHER	2
2	CP4350-106	M6 x 14 SCREW, COUNTERSUNK	6
3	CP5119-134	PAD RETAINING R CLIP	1
4	CP5200-173	M10 x 1.50 SOCKET HEAD CAP SCREW	2
5	CP5427-101	M10 NYLOCK NUT	2
6	CP5500-105	BALANCE BAR	1
7	CP5500-106	SPHERICAL BEARING	1
8	cp5500-113M	SCREW M4 x 5	2
9	CP5500-124	WASHER	6
10	CP5500-133	STUD M8x1.25, 28.0 LONG	2
11	CP5500-148	SNAP RING	1
12	CP5500-150 (MODIFIED)	RUBBER BOOT	1
13	CP5500-150	RUBBER BOOT	1
14	cp5500-312	CLEVIS - METRIC	3
15	CP5500-318	FOOTPAD	3
16	CP5500-321	TORSION SPRING	1
17	CP5500-322	SPRING SUPPORT BUSH	1
18	CP5500-323	CLUTCH CLEVIS PIVOT RETAINER	1
19	CP5500-324	M6 SCREW x 30, FLAT GRUB	1
20	CP5007-131	GLYD RING BEARING	1
21	cp5516-127	SPACER	5
22	CP5500-318	M6 SCREW, 6M SHOULDER	1
23	ME21002	M6 LOCK NUT	1
24	cp5500-108	BARREL NUT	2
25	CP5500-41	CLUTCH PEDAL SUB-ASSY	1
26	CP5500-40	BRAKE PEDAL SUB-ASSY	1
27	CP5500-42	THROTTLE PEDAL SUB-ASSY	1
28	CP5500-329	THROTTLE GUARD	1
29	CP4296-116	M10 x 1.50 SOCKET HEAD CAP SCREW	1
30	CP5500-331	M10 x 1.50 SOCKET HEAD CAP SCREW	1
31	CP5500-134	M6 x 1.00 HEX BOLT	2
32	ME21001	M6 x 1.00 NUT	1
33	cp5500-340	BASE PLATE	1
34	CP5500-338	METRIC CLUTCH CLEVIS	1
35	CP5500-314	M6 SCREW x 40, FLAT GRUB	1
36	cp5500-345	CLEVIS - UNF	2

FOR UNF VERSION - CLEVIS - IMPERIAL
ITEM 14 - CP5500-191 - CLEVIS - IMPERIAL
ITEM 34 - CP5500-336 - CLUTCH CLEVIS - IMPERIAL

AP Racing
Wheler Road
Coventry
CV3 4LB

TITLE: GENERIC FLOOR MTG PEDAL BOX
DRG NO.: CP5500-505CD
SCALE: 1:1
SHEET 2 OF 3
DRAWN: Chris Annesworth

A0 INSTALLATION DRAWING

FIRST ANGLE PROJECTION

OPTIONAL EXTRAS

THROTTLE BELLCLASSIS KIT
PART NUMBER: CP5500-044
SOLD SEPARATELY

THROTTLE LINKAGE KIT
PART NUMBER: CP5500-043
SOLD SEPARATELY

LARGER FOOTPAD
PART NUMBER: CP5500-330
SOLD SEPARATELY

ADJUSTER COULD BE MODIFIED TO SUIT CABLE ANGLES.

R82.50 (1.404 RATIO)
R53.50 (1.202 RATIO)
R44.50 (1.000 RATIO)
R35.50 (0.797 RATIO)
R26.50 (0.595 RATIO)
R44.50
R37.00
R80.00
63.00
83.00
35.00
7.00
2.00
70.0 - 61.0
M6x1.00 THREAD

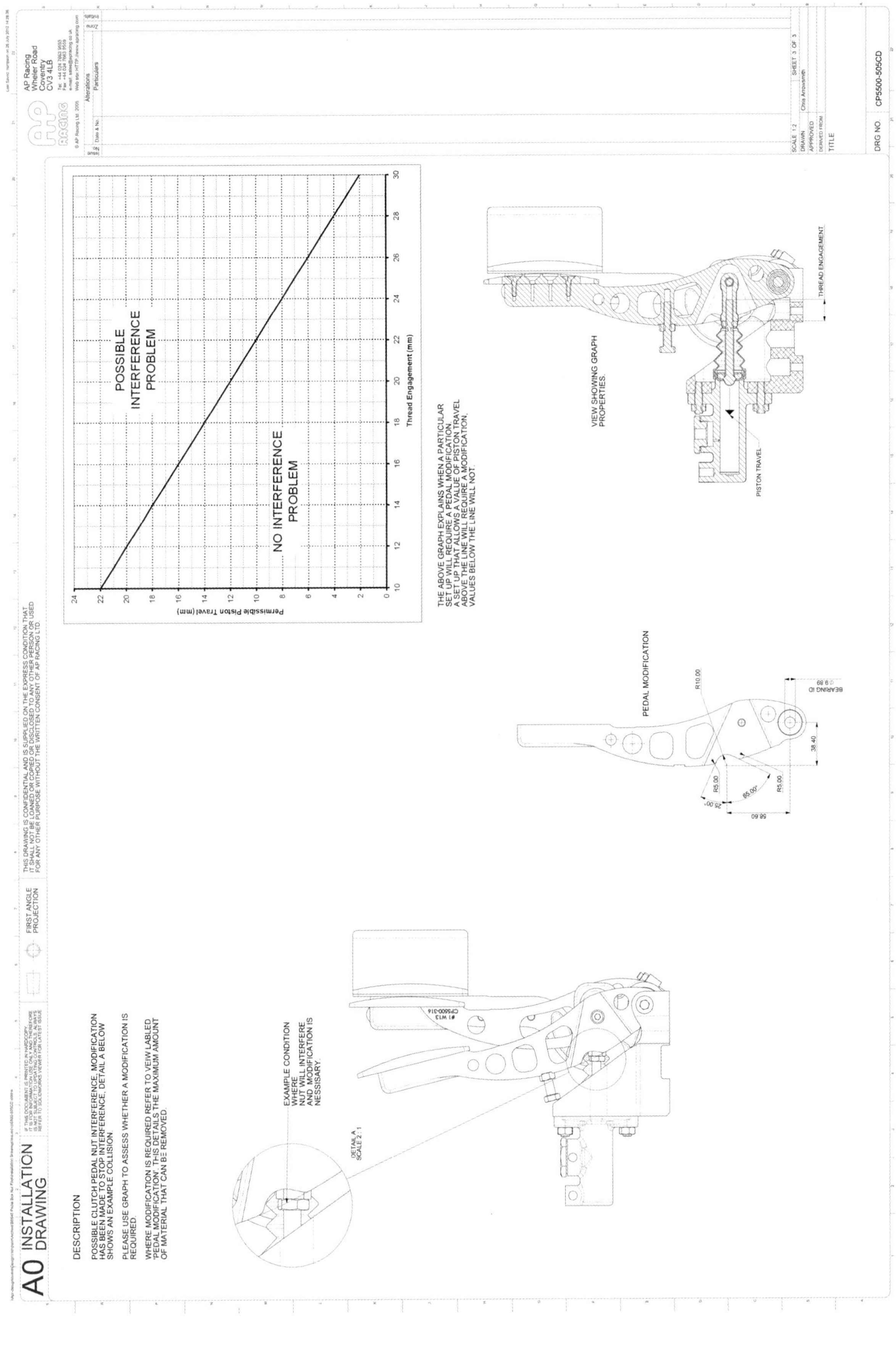

Brakes 113

Figure 3.45 Pedal box

It is for this reason that the braking system of a vehicle is designed to distribute the braking pressure with a bias to the front wheels. This helps prevent the traction limit of the rear tyres from being exceeded and hence locking the wheels, while being able to take advantage of the increased traction capacity of the front. It must be remembered that in wet or low-grip conditions, braking power (g) will be reduced and, therefore, so will weight transfer. As a result the brake bias will need to be adjusted rearward to avoid locking up the front wheels. Some cars can use a restrictor in the rear brake line to balance the brakes instead of the pedal-box-based bias bar.

The choice of MC is important and must be correctly paired to the choice of callipers, since the cross-sectional bore area has a direct relationship to the pressure in the braking system and, combined with the mechanical advantage offered by the pedal itself, will have a direct relationship to the system feel.

Since pressure, force and area are linked by the equation:

$$\text{Pressure} = \frac{\text{Force}}{\text{Area}}$$

then the force applied to a brake disc is determined by the pressure applied by the driver (including the mechanical advantage offered by the pedal ratio) and the areas of the MC and calliper pistons in the following equation:

$$\text{Force on rotor} = \text{Pressure of MC} \times \text{Area of MC} \times \frac{\text{Area of calliper}}{\text{Area of MC}}$$

Therefore, an increase in MC bore size has an inverse effect on the hydraulic pressure present in the system. However, the fluid displaced by the MC is directly proportional and hence a smaller diameter cylinder will require greater pedal movement in order to achieve the same displacement. This must always be considered when alterations to a system are being made.

3.4.3 Callipers

The brake callipers must be chosen correctly for the application and there are many different types of calliper on the market. They vary in construction, material, number and diameter of pistons. The ultimate callipers are of monobloc construction, with multiple pistons of unequal diameter. These callipers are used in top-level motorsport such as Formula 1 and Le Mans Series (LMS). Quick release callipers are also used in LMS racing so that the brake pads can be quickly removed and new ones refitted if needed in a very short time. They offer many advantages over more common two-piece callipers with an equal piston diameter.

In order to provide the best brake system operation and feel, there must be minimum flexing of the callipers as this leads to an imprecise and spongy brake pedal. A monobloc calliper is machined from just that – a single block of material without a joint line; in the case of Formula 1 this is a lithium aluminium alloy. Any flex in the calliper results in the lower part opening up, causing radial tapered pad wear.

The pistons fitted into such a calliper would be of different sizes along the length of it. If we consider again the Formula 1 calliper, we would see three pistons per side (the limit imposed by regulations), as shown in Figure 3.47. Multiple pistons allow for larger pads to be used and for even pressure to be applied to the backing plate.

The reason for different-sized calliper pistons is to achieve even and consistent pad pressure. The leading edge of a brake pad wears faster than the trailing edge; this is because the leading edge of the pad is drawn down into the rotor surface by the friction couple when the brakes are applied while the trailing edge is lifted. Particles of friction material are also carried from the leading to trailing edge of the pad. In effect, the trailing portion of the pad is riding on this layer of friction material. By providing an optimally designed larger calliper piston at the trailing edge of the pad, wear can be evened out along the length of the pad.

Uneven pad wear can often occur with single-sized multi-piston callipers. This can result

Figure 3.46 Monobloc and two-piece callipers

in the brake pedal gradually becoming less responsive with increased pedal travel over the course of the pad's life.

The piston material is another factor that can separate the top-level brake calliper from the entry or lower-level brake calliper. In Formula 1, the material of choice is titanium. This has a number of advantages that outweigh the inherent extra cost in sourcing and machining such an exotic material. It is extremely lightweight, thus aiding the continuing goal to reduce unsprung weight. Furthermore, it has better thermal insulation properties than equivalent steel. This helps to keep the heat generated by the brakes away from the fluid in the system, reducing the risk of fluid boiling and bubbles forming.

Callipers have two different mounting types: radial- or lug-mounted. Radial-mounted callipers offer a more rigid mounting system than a lug-mounted system and the radial system can also cope with rotor diameter changes by using a simple spacer. The two are mounted differently: in the radial mount the fixing bolts go down through the top of the calliper and into the upright; in the lug mount two lugs are machined into the shape of the calliper, with the bolts passing through the front face of the calliper and into the side of the upright.

Figure 3.47 Multiple-sized pistons in a calliper

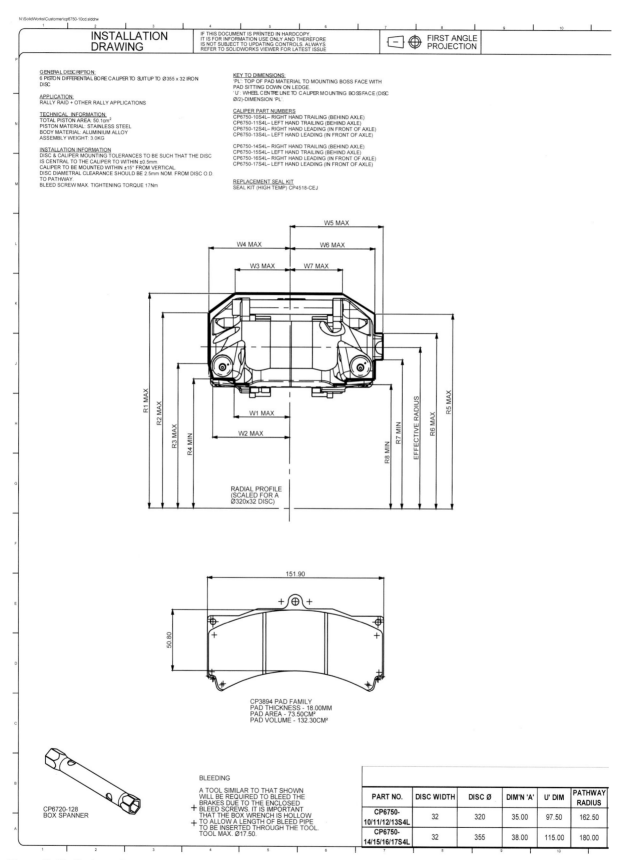

Figure 3.48 Brake caliper

Brakes

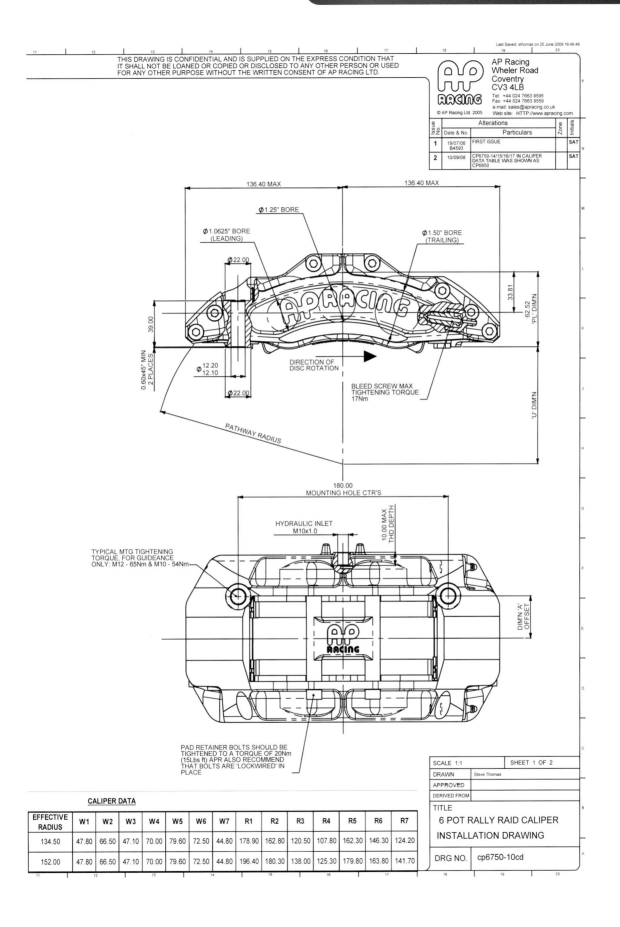

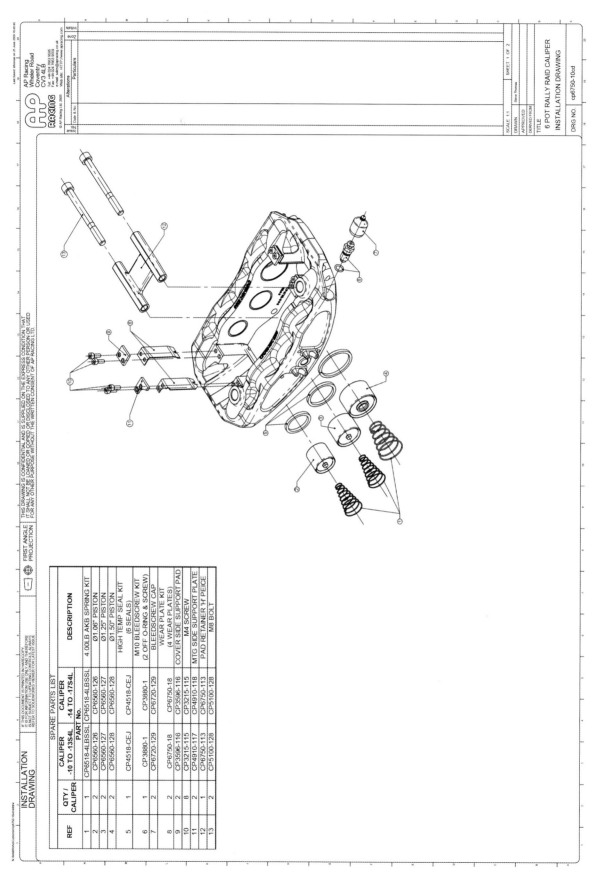

Figure 3.49 Brake caliper

3.4.4 Discs

The brake discs (also called 'rotors') are one of the most important parts of the braking system. As with the calliper, the correct selection of brake disc is very important, with many different types, sizes, styles and materials available – each with their own applications, advantages and disadvantages. Disc material includes steel, cast iron and carbon.

Drilled or grooved

A passenger road car would typically be fitted with discs with a plain surface. However, in motorsport or high-performance applications the disc surface can often be found to be drilled or grooved or both. The drilled brake disc has a series of holes drilled around its face to allow the gases generated during braking to escape. This prevents a build-up of such gases, which can result in the pads 'floating' and not functioning properly. However, these holes introduce stress points to the discs, which can propagate cracks and disc failure in extreme environments.

The best option is to have radial grooves (or slots) machined into the disc face, as in Figure 3.50 (right). Much like the drilled brake disc, the slots provide an escape for the hot gases, but unlike the drilled holes, high stress points are not introduced into the disc. This makes the grooved disc the favourable option for high-performance and race car applications.

Solid or vented

Brake discs are available in either a solid form, or with a vented section between the two faces. Although there is no direct relation to braking performance (with regards to clamping force or energy generated) for a like-for-like vented and solid rotor, the selection must be carefully considered and a number of vehicle factors must be taken into account. Typically, the solid disc would present less weight than the vented disc. However, a vented disc is able to dissipate heat more quickly and has a higher heat capacity. This means that selection must be based on the mass of the vehicle and the cooling available to the brakes themselves. Since rear bias is often less than that of the front, then the requirement to dissipate heat can also be less, and it is not uncommon to find solid discs on the rear of a car fitted with vented discs to the front.

Two-piece and floating discs

Two-piece discs are often seen in high-performance and motorsport applications. These use an aluminium central bell to which the disc itself is mounted. The disc material can often be mounted in a 'floating' fashion – that is, having a small amount of free play designed into its mountings. This set-up can present a number of important advantages for the race car. Firstly, the aluminium central bell

Figure 3.50 Grooved and drilled (left) and grooved (right) discs

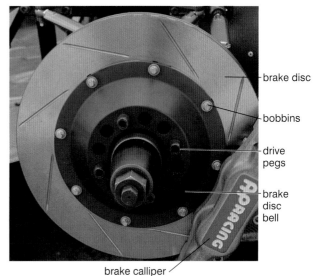

Figure 3.51 Floating disc assembly

presents less weight than a one-piece disc. This weight is unsprung and rotational so, by reducing it, there can be an improvement in handling response and acceleration. Secondly, the brake disc must experience large temperature fluctuations, which may warp or distort a one-piece disc, giving pedal vibration or knockback of the pads. By using a two-piece floating disc, the temperature is spread more evenly around the material and the float in the mountings allows for thermal expansion.

3.4.5 Brake pads

Brake pads are another critical choice for the car and are available in materials such as carbon, carbon metallic, and cerametallic. A brake pad is normally chosen on the following principles:

- Temperature vs. Cf curve – this shows the pad's friction level across a temperature range. Some pads may have a high Cf rating but may work poorly in cold conditions or very hot conditions.
- Wear rate – this shows how quickly the pad will wear out; race durations and budget will determine the outcome.
- Bite and release characteristics – this is how the pad reacts initially after being forced against the disc and how it is released from the disc. The choice is down to feel and balance but can also be based on the rest of the brake system design. For example, a brake system set-up that offers little modulation would not suit a pad with a harsh bite and release characteristic.

Pad fade can occur when the maximum operating temperature of the pad is exceeded. The brake pedal will remain firm but the stopping power will be significantly reduced, so the pads will need to be brought back down to temperature before they can work efficiently again.

Pad glazing occurs when the surface of the brake pad becomes lightly polished due to being used at a low temperature with a low pressure applied – this significantly reduces friction. They can be deglazed with a low-grit paper until they become a matte colour.

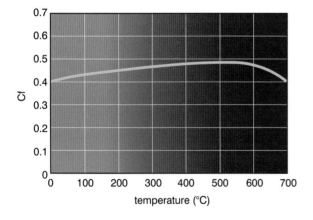
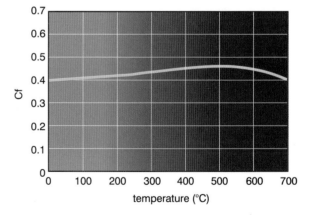

Figure 3.52 Two different brake pad specifications

Always follow manufacturers' procedures for bedding in pads in order to get the best out of them. Never start a race on new pads!

3.4.6 Brake lines

The brake lines on a race car differ from those on a typical passenger road car. In the search for the ultimate pedal feel and response, the rubber flexible lines are replaced with stainless steel braided lines. These resist the expansion that happens to rubber lines due to the high line pressures during braking and provide the driver with a firmer pedal. The stainless steel braiding also provides greater strength to the line and an increased resistance to abrasion.

3.4.7 Fluid

The brake fluid is incompressible and ensures a firm pedal feel. This hygroscopic fluid absorbs moisture over time, which reduces the boiling point. Brake fluid is rated at a dry (without water content) and wet (with water content) boiling point. It is important to ensure that neither limit is reached as air bubbles will form in the lines, which will then become compressible, resulting in a soft and spongy pedal and poor braking. Changing fluid regularly will ensure brake pedal consistency. The quality of fluid varies and it must never be stored in a lidless pot as moisture will find its way in. Castrol SRF is a commonly used brake fluid in motorsport; it has a dry boiling point of 320 °C and a wet boiling point of 270 °C, which are higher than most other brands.

3.4.8 Cooling and data logging

Brake ducts are commonly used to cool the front brake assemblies (and sometimes the rear brakes), and should be used to keep the brakes within their optimum working temperature. They are also used to add additional downforce. You must ensure that they are not overcooled, as blanking the ducts or using smaller inlets reduces the cooling effect during cool or wet conditions, or undercooled, which can cause pad fade or warping. Infrared sensors and brake disc temperature paint can be used to monitor brake temperature in order to ensure that brakes are kept within an optimum temperature range.

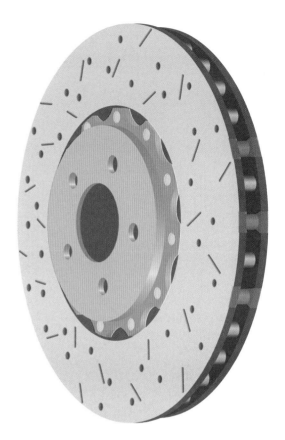

Figure 3.53 Brake disc temperature paint: green changes to white at 430 °C, orange changes to buff at 560 °C, red changes to white at 610 °C

3.4.9 Pad knockback

Pad knockback occurs when the brake pad is forced away from the disc while the vehicle is in a dynamic state. This will then lead to a soft brake pedal, which on initial depression will produce an inconsistent amount of pedal travel and can be dangerous for the driver. If the driver can tap the brake pedal and then get a better performance when they press it again, this is most likely pad knockback/off. This cannot always be avoided and both one- and two-piece disc set-ups can be affected by this scenario. Cars clattering off kerbs are the main cause of pad knockback, while too much or too little lateral float in floating discs can also affect it. While the brake system can cause knockback, other areas that can be problematic, if not designed correctly, include the hub/upright assembly flexing (or deflecting) and caliper to pad to disc alignment with temperature change.

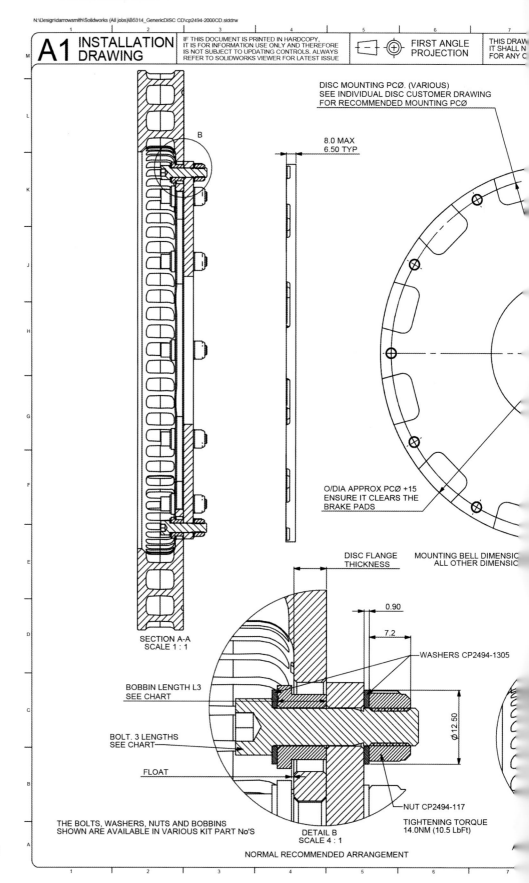

Figure 3.54 Floating disc assembly

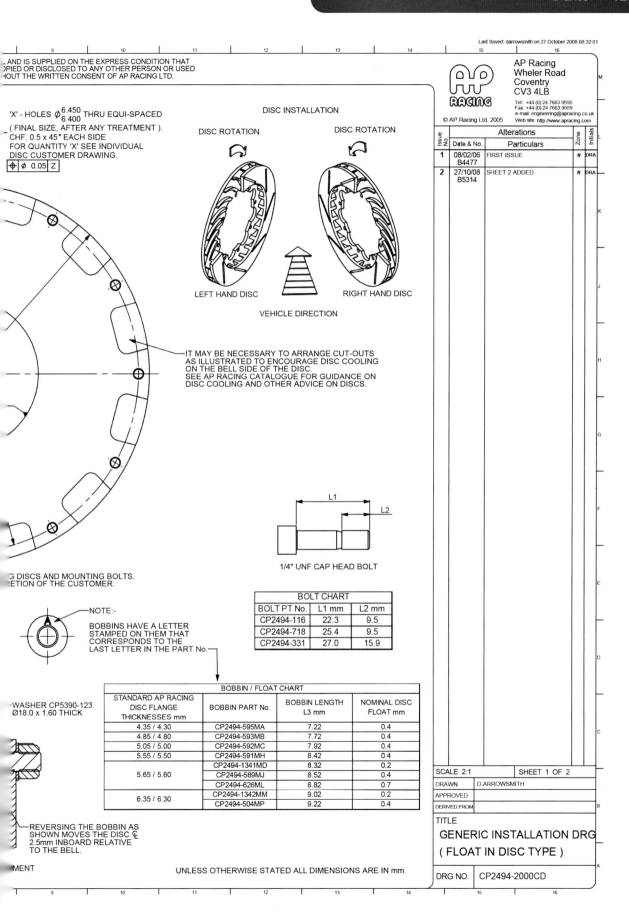

3.4.10 Basic calculations

Some simple calculations can be applied to the braking system so that we can visualise what each part of the brake system does:

$$\text{Pedal ratio} = \frac{\text{Distance from pedal pivot to the centre of the footpad}}{\text{Distance from pivot point to MC}}$$

$$\text{Brake line pressure (psi)} = \frac{\text{Foot force (lb)} \times \text{pedal ratio}}{\text{Area of MC (in}^2\text{)}}$$

$$\text{Clamping force of calliper (lb clamping force)} = \text{Brake line pressure (psi)} \times \text{Total piston area (in}^2\text{)}$$

$$\text{Braking torque (lb-ft)} = \frac{\text{Effective disc radius (in)} \times \text{Clamping force (lb)} \times \text{Cf of pad against disc}}{12}$$

$$\text{Effective disc radius (in)} = \frac{\text{Disc useable outside diameter (in)} + \text{Disc useable inside diameter (in)}}{4}$$

CHAPTER 4

Aerodynamics

This chapter will outline:

- Principles
- Body shape
- Wings
- Underfloor and diffusers
- Additional parts

4.1 Principles

Aerodynamics is the study of the way in which a body flows through a fluid. This is how the car or one of its parts flows through the air. Aerospace and automotive industries (or more specifically, motorsport) have developed aerodynamics for similar reasons over the years but for different outcomes.

Aerodynamics: study of a fluid (air in this case) with a body (car part or component) moving through it.

4.1.1 Downforce and drag

Downforce and drag are the two most important parts of aerodynamics and are closely related. We want lots of the first one, but we do not want the second one at all. In aerospace, downforce is substituted for lift.

Downforce is the force of air pushing down on the car due to the way in which the air is being worked.

Drag is a longitudinal force that faces the car head-on and saps horsepower at any velocity – the higher the velocity, the more horsepower is used to overcome drag.

$$\text{Aerodynamic efficiency} = \frac{\text{Downforce}}{\text{Drag}}$$

So the higher the number, the better the efficiency.

Example:

$$\frac{500\,\text{N downforce}}{250\,\text{N drag}} = 2.0$$

$$\frac{500\,\text{N downforce}}{125\,\text{N drag}} = 4.0$$

Aerodynamic balance is also key – a car with an imbalance of downforce will have understeer or oversteer in high-speed corners, making control of the car difficult for the driver. From considering where the downforce is applied along the car, we also need to consider the car's **centre of pressure**.

Centre of pressure: the point at which the aerodynamic forces are balanced. Ideally this should be in close proximity to the centre of gravity (CG) in order to provide a stable car in high- and low-speed scenarios.

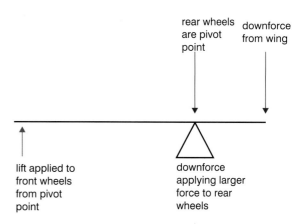

Figure 4.1 Beam diagram showing potential downforce effects

Downforce is very advantageous as it adds vertical tyre load but without the penalty of increasing the car's mass.

4.1.2 Bernoulli's theory

The key principle (known as Bernoulli's theory) is to have a difference of speed over the top and bottom surfaces of the car or component. As speed increases, pressure reduces, so the aim is to have very low pressures underneath the car body and other aerodynamic aids, and then a higher pressure on top. The car is effectively being sucked and squashed into the floor, creating downforce and, therefore, grip (see Chapter 3 Chassis to find out more about vertical load).

In most cases, any amount of downforce comes with a penalty – drag – so designers and engineers aim to produce a car with maximum downforce with the minimum amount of drag.

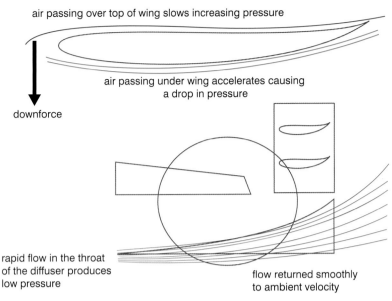

Figure 4.2 Wing and underfloor airflow

Downforce = $0.5 \times \rho \times V^2 \times Cl \times A$

Drag = $0.5 \times \rho \times V^2 \times Cd \times A$

Where:

ρ = Air density (kg/m³)

V^2 = Velocity (m/s)

Cd = **Coefficient of drag**

Cl = **Coefficient of lift**

A = Frontal area (m²)

Note: when considering drag, A could be used for plan area as well as frontal area.

Example:

If Cl = 3, Cd = 0.6, A = 1.7 m², ρ = 1.225 kg/m³, V = 50 m/s, we can calculate:

Downforce = $0.5 \times 1.225 \times 50^2 \times 3 \times 1.7$
= 7809.4 N (to 1 d.p.)

Drag = $0.5 \times 1.225 \times 50^2 \times 0.6 \times 1.7$
= 1561.9 N (to 1 d.p.)

To obtain a weight equivalent, divide these numbers by 9.81 m/s² (the acceleration of gravity) which will give you a figure in kilograms.

Coefficient of drag (Cd): a unitless number that allows comparison of drag incurred by different-sized and shaped bodies.

Coefficient of lift (Cl): in this case, downforce, this is a unitless number that allows comparison of downforce (negative lift) incurred by different-sized and shaped bodies.

We can see from the equation opposite that drag and downforce are functions of speed. So as a car slows under braking, we should expect the driver to be able to brake harder initially and then be able to bleed the brake pressure off as the aerodynamic downforce reduces. Referring back to our tyre G-plot in Chapter 3 Chassis, the faster we go, the more downforce the car should create (by a squared function of velocity), so we should be able to increase our performance in braking, accelerating and cornering. Different disciplines of motorsport will use varying aerodynamic set-ups, for example, if you compare a hill-climb single-seater, that has an aggressive aerodynamic set-up (due to low average speeds) to a Grand Prix (GP2) circuit car, you can see how different the configurations can be.

Figure 4.3 Gould GR55b

Figure 4.4 GP2 – notice the difference in the wing design compared to Figure 4.3

4.1.3 Air density and body shape

Air density has a direct effect on performance: the denser the air, the more drag and downforce there is. Sadly, this is a factor we cannot control, although it is good to understand this and log it during a race to avoid any confusion with previous data when the car was faster or slower (see Chapter 1 Engine for additional information on air density and other ambient conditions).

The frontal area has a huge bearing on the drag and downforce of the car and should be minimised as much as possible. The most noticeable issue with the frontal area of single-seaters is the tyres, which often take up 30–40 per cent of the frontal area but, of course, these cannot be removed!

The coefficients of drag and downforce are based on the car's body shape (under and above) and the aerodynamic parts added to it. Increasing these areas has a dramatic effect on aerodynamic performance.

Getting the right balance of aerodynamics is important. However, there will be different conditions at every circuit and aerodynamic performance changes as conditions change. Too little downforce and the car will be slow in high-speed cornering; too much downforce and the car will be slow on the straights. The right aerodynamic balance between the front and rear of the car is needed for the car to be driveable through each corner entry, apex and exit.

4.1.4 Data logging

Data logging can be very helpful as analysing information, such as throttle histograms and speed traces, can reveal whether the circuit is the type where aerodynamic downforce or drag will have the most effect on the lap time. A circuit with a high average speed, but also one with a high percentage of the lap with full throttle would potentially show that reducing drag is the most important objective. In the UK, circuits that require low drag include Silverstone and Castle Combe, while examples of high downforce circuits are Cadwell Park and Lydden Hill.

4.1.5 Measuring aerodynamics

Aerodynamics is analysed and measured in the following three principle ways.

Wind tunnel testing

This is the most trusted and most accurate form of testing available, as conditions can be controlled and scale parts and cars can be efficiently tested at numerous speeds, yaw angles and other parameters. The disadvantage is that they cost a huge amount to run (or to be rented daily), and are very labour intensive.

The beauty of wind tunnels (and computational fluid dynamics to some extent) is that you can control atmospheric conditions, roll, yaw, pitch, steering angle and other dynamic motions. Wind tunnels can also have a flat rolling road fitted to

Figure 4.5 Using a smoke stream when testing in a wind tunnel

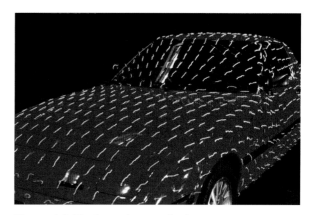

Figure 4.6 Testing using wool tufts

simulate wheels revolving at the correct speed (which disrupts airflow) as the fan is pulling the air across the car. To help visualise flow in the wind tunnel, you can use smoke streams and wool tufts.

Computational fluid dynamics (CFD)

This can test parts and cars prior to final manufacturing. Multiple runs and simulations can be carried out reasonably quickly, however, processing time depends on the power of the computers and how accurate the test needs to be. Visualisations of pressures and speeds can be easy to view and, as with wind tunnels, different conditions and motions can be controlled. CFD is constantly being developed – it does not currently provide the most reliable results, although it does give an insight into aerodynamic performance.

Aerodynamics

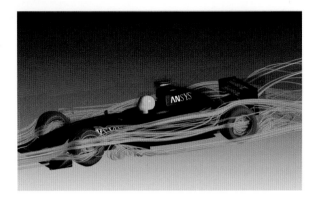

Track testing

The beauty of track testing is that it is real. However, collecting data in track conditions can be difficult and full-size and full-strength components must be created. On circuits, flo-viz paint is often used to visualise airflow.

Figure 4.7 Computational fluid dynamics (CFD) testing

Figure 4.8 Track testing

Figure 4.9 Using flo-viz paint while track testing

4.1.6 What affects aerodynamics?

The aerodynamics of a car can be affected by many things:

- Wind speed and direction – this can change the car's top speed, braking ability and chassis balance as the car completes laps in different directions. Wind speed can be easily measured with a pitot tube fitted at the front of the car.
- Air pressure and humidity – a higher air density increases the fluid that the car has to work through, giving more downforce and more drag, but also possibly more power as the engine is feeding on a greater mass of air.
- Air and track temperature – air temperature affects air pressure, while track temperature can make it difficult to balance tyre temperatures and, therefore, aerodynamic balance.
- Weather conditions – the choice of downforce vs. drag will often be different between dry and wet conditions.

Some areas of aerodynamics are largely fixed by design, such as the under- and over-body of the car, while others are made adjustable for trimming purposes, such as wings.

> Some useful terms:
>
> **Boundary layer:** a thin layer of static or slow airflow next to the surface it is flowing over. The air is slow due to surface friction.
>
> **Dynamic pressure:** movement energy of the air (or fluid) expressed as $0.5 \times \rho \times V^2$.
>
> **Induced drag:** drag caused directly by vehicle body shape and layout.
>
> **Stagnation point:** the point, usually at the front of the body, where the air velocity is zero and air pressure is high.
>
> **Surface drag:** drag caused by the friction of the body's surface.

We can also use some calculations to help us understand the principles of aerodynamics and how it can affect the performance of a race car.

> Brake horsepower (bhp) absorbed by drag
>
> $$= \frac{C_d \times A \times V^3}{1225}$$
>
> Where:
>
> C_d = Coefficient of drag
>
> A = Frontal area (m²)
>
> V = Velocity (m/s)
>
> So taking our data from earlier, we can calculate the following:
>
> $$\frac{0.6 \times 1.7 \times 50^3}{1225}$$
>
> = 104 bhp (to the nearest whole number)

> Maximum cornering speed = $\sqrt{\dfrac{\mu \times R \times r}{M}}$
>
> Where:
>
> μ = Tyre coefficient of friction with the ground
>
> R = Car mass plus downforce (kg)
>
> r = Corner radius (m)
>
> M = Cars mass (kg)
>
> So, we can then calculate maximum cornering speed by assuming μ = 1.4, r = 50, R = 1500 (or 1000) and M = 1000:
>
> $$\sqrt{\frac{1.4 \times 1500 \times 50}{1000}} = 10.25 \, \text{m/s}$$
> $$= 22.9 \, \text{mph (to 1 d.p.)}$$
>
> (with 500 kg downforce)
>
> $$\sqrt{\frac{1.4 \times 1000 \times 50}{1000}} = 8.37 \, \text{m/s}$$
> $$= 18.7 \, \text{mph (to 1 d.p.)}$$
>
> (with no downforce)

From the calculation, you can then use the following equation to work out how much time can be saved through a corner by adding downforce, once you know the angle of the corner:

$$\text{Time taken} = \frac{\text{Distance}}{\text{Speed}}$$

Example:

A corner of 90° and radius of 50 m (using $2\pi r \div 4$) would give you a corner distance of 78.5 m.

$$\frac{78.5}{10.25} = 7.7\,\text{s} \quad \text{(with downforce)}$$

$$\frac{78.5}{8.37} = 9.4\,\text{s} \quad \text{(no downforce)}$$

Taking this time saving of 1.7 s over one corner, you can realise how much difference downforce would make over a whole lap!

4.1.7 What does dirty air mean?

You often hear this term used in TV commentaries of Formula 1 and Le Mans Prototype (LMP) racing. It means that instead of the car running into a **freestream** of air, the car in front is affecting the airflow before the car behind hits it. As a consequence, we often find that the car behind will lose downforce and the aerodynamic package will not be efficient in turbulent conditions.

> **Freestream:** airflow around the body, which is undisturbed by the body moving through it.

A related term, aerodynamic 'tow', is when the car in front is cutting through the air and producing a low-pressure and turbulent area directly behind it. The advantage for the car behind is reduced drag, and, therefore, better acceleration. Caution must be taken, however, to avoid losing downforce and making an error; elevated engine and brake temperatures can also occur from inadequate cooling whilst in the tow.

> **Turbulent flow:** flow that begins to swirl or mix while flowing around the body.

4.2 Body shape

Airflow under and over the whole body of the car is as important as the aerodynamic aids that you see on a modern race car.

The aerodynamic performance of the car is very sensitive to the front and rear ride height of the car, which is why Formula 1, LMP and similar cars have very little suspension movement.

As the ride height at either end of the car changes from pitching and roll effects, or a combination of the two, the aerodynamics will change, in particular the downforce and aerodynamic balance.

Most race cars also use 'rake', which is when the car has a higher ride height at the rear than the front. This is often to provide further aerodynamic benefits, which will be discussed later.

All cars have a sensitivity to ride height change and roll angle when considering the aerodynamic performance. Aerodynamic maps can be produced using information about the working range of the ride heights and roll angles during a lap of a specific circuit. These can then be used to aid testing via a wind tunnel or CFD. From there, an aerodynamics engineer can work out the optimum range of ride heights for the car in order for it to produce the most consistent aerodynamic balance and the best efficiency. You can then also test the wing angles and other add-ons, such as dive planes and gurney flaps, to see how the aerodynamic map is affected by these changes. Then, for example, you can be sure that if you need to add total downforce while keeping the same aerodynamic balance, you know how much wing needs to be added at each end of the car. When considering roll angle, this will help to determine the correct spring and anti-roll bar rates to run (see Chapter 2 Chassis for more information).

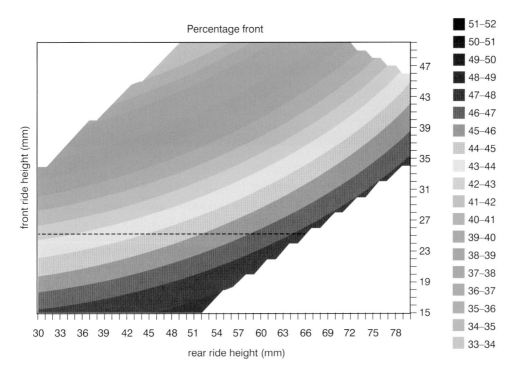

Figure 4.10 Aerodynamic maps

Rapid changes in aerodynamic balance make the car difficult to drive and to set up. The map will show the importance of ride height and the car's yaw angle. As discussed in Chapter 2 Chassis, ways of ensuring ride height stability could include preload and anti-dive/squat. Keeping the aerodynamic balance stable during braking, accelerating and cornering will help stability and make the car easier to drive – making it easier to drive will make it faster.

The designer of the car has to have some key objectives in the design of the body shape, which mainly depends on CdA and ClA (coefficient of drag, downforce and frontal area).

A quick look at Formula 1 cars over the years will show how much the body shape of the cars have changed to include high noses, carefully sculptured sidepods and additional guide curves and vanes to help drive air to the right areas and away from the tyres. Compare the two cars in Figure 4.11 to identify the key changes to body shape over the years.

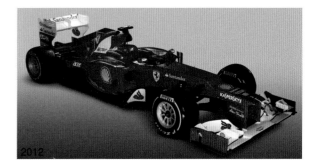

Figure 4.11 Formula 1 body shape development

4.3 Wings

Front and rear wings are major producers of downforce on a car. Thus, a sound understanding of their operation is important.

4.3.1 Wings

The wing (or aerofoil) can be situated at the front or back of the car, often both, and is shaped to produce downforce. The basic principle of its operation is to produce downforce using the shape of the upper and lower surfaces. The air flowing over the top travels a shorter distance, increasing pressure. In order to keep up, the air travelling on the underside has to travel a larger distance, and has to accelerate so it meets the air on top, at the **trailing edge** of the wing, resulting in low pressure underneath the wing. High pressure on top and low pressure underneath causes downforce.

The profile (shape) of the wing determines many of its characteristics in terms of downforce and drag levels. However, we also need to look at the **angles of attack** (AoA) that the wing can run within.

Thousands of wing profiles were designed and tested many years ago by the National Advisory Committee for Aeronautics (NACA), which is the old name for NASA. These tried and tested profiles can be easily adapted to suit a motorsport requirement. Remember that the principles in aeronautics are essentially the same as motorsport principles, just upside down!

Wings develop downforce based on parameters, such as how many elements are used, their dimensions (including **span**, **chord**, camber and thickness) and what AoA they are being run at.

The wing can generate downforce at very small AoAs because the camber of the wing itself can provide the difference in air velocity around the upper and lower surfaces, without any AoA. However, when the AoA is increased, there tends to be an increase in the low pressure underneath and an increase in the high pressure on top – this also creates drag with the high pressure and marginal increase in frontal area. Every wing will have an individual maximum AoA that it can operate at. As the AoA increases, the air which is moving close to the surface (**attached flow**) begins to move away. Starting

Angle of attack (AoA): angle between the chord line and the freestream air.

Aspect ratio: calculated by Span ÷ Chord.

Camber line: race car wings need to have camber in order to optimise the car for efficiency. The camber means that the lower surface is more curved than the upper surface. The camber line is, therefore, always an equal distance from the upper and lower surfaces from the leading to trailing edge.

Chord: distance from leading to trailing edge.

Leading edge: front tip of the wing.

Span: width of the wing from one end to the other.

Trailing edge: rear tip of the wing.

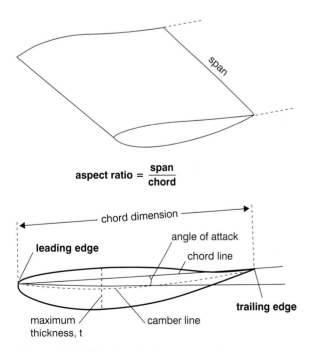

Figure 4.12 Wing terminology

from the trailing edge underside, this will work its way further forward as AoA is increased until the wing stalls. This reduces the wing's ability to produce downforce and increases drag, with the wing stalling from the **separated flow**.

Attached flow: (or laminar flow) where airflow follows the surface of the body it is flowing around.

Separated flow: airflow that no longer follows the shape of the body.

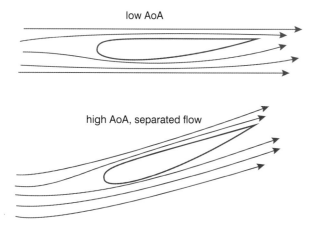

Figure 4.13 Attached and separated flow examples

A larger camber angle is expected to produce higher amounts of downforce but consequently can also produce more drag, so flow separation will occur at a higher AoA.

A single element only has the ability to produce a certain amount of downforce before flow separation occurs. An option to increase downforce on this is to add another element and create a biplane wing or multi-element wing, or add even more elements to produce more downforce. A biplane wing can have a flap added behind the trailing edge to supplement more downforce. Adding a flap increases the plan area of the total wing, giving more downforce, while also increasing the camber angle, also giving more downforce. You will also notice that the AoA is increased on this type of wing layout. The positioning of the flap in relation to the wing is vitally important to its performance. There needs to be a form of overlap and slot gap between the main plane trailing edge and leading edge of the flap – approximately a 2 per cent gap and 4 per cent overlap of the chord length. The convergent slot gap is to allow for some of the high-pressure air on top to bleed on to the lower surface and help to energise the underside of the wing to prevent flow separation.

More flaps and slats can be added to a wing assembly to further increase downforce and efficiency levels.

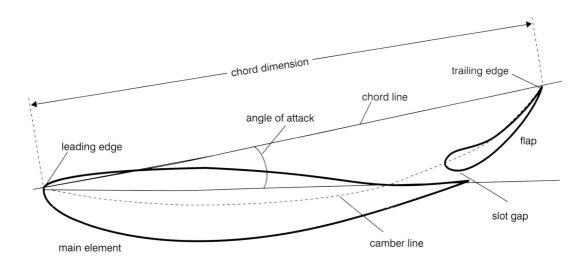

Figure 4.14 Multi-element wing terminology

Twisted wing elements

Some cars also use twisted wing elements to maximise the aerodynamic efficiency. In most cases, by the time the freestream of air has worked its way over the body, it is flowing at different angles across the vehicle's width. So, along the wingspan, we can twist the profile (symmetrically) so that we can ensure that the wing's performance is optimised across its whole span.

Figure 4.15 Twisted wing profile

Gurney flaps

A gurney flap is a small right-angled strip that is attached to the rearmost elements' trailing edge and is used to produce additional downforce. It forces the air upwards and also slows down flow in front of it, increasing **static pressure**. The gurney flap (when kept relatively short) provides a useful downforce increase with little drag penalty. With some wings, it can be used as a tuning tool as it can be removed or changed for a different size (height or span) or shape.

> **Static pressure:** ambient pressure present within a certain area.

Figure 4.16 Gurney flap, is the small vertical grey part at the back

End plates

These provide a way of ensuring that the high-pressure air stays on the upper surface and cannot migrate to the lower pressure underneath. They also provide an added increase in downforce and reduction in drag (although very large end plates can produce unwanted instability if crosswinds on a circuit hit a large area of flat end plate). By using end plates, the spillage of air pressure is reduced, which allows the effective aspect ratio of the wing to be increased. The **wing tip vortices** coming off the wing are reduced (starting after the wing, not before) and are guided straight back instead of

Figure 4.17 Wing end plates

producing a wake behind the car and inducing drag. Wings often tend to be positioned at the maximum height that regulations allow in order to be in a freestream of air that is not affected by the rest of the car. This means there may not be much scope for end plate above the wing, but we can use a lot of end plate below the wing to stop migration.

> **Wing tip vortices:** air that begins to rotate as it spills off the sides of the wings, unless controlled. It can cause heavy amounts of drag and reduce efficiency.

The mountings for a wing must be stiff in order for the wing not to move with the huge forces that will be applied to it. Some race series, such as Formula 1, have wing pullback tests to ensure that wings do not flex to gain an advantage through low drag at high speeds. Some wings are mounted via the end plates to the body, while others are mounted centrally from a pair of stays, often mounted to the gearbox casing. These must be aerodynamic in themselves in order to reduce drag and so not disturb flow to the wing.

centre mount

upper elements supported by end plates

Figure 4.18 Wing mountings

low Stohr wing

high Ligier wing

Figure 4.19 High- and low-positioned wings

Adjustment of the wing must be quick and easy; most have bonded threaded inserts in the wing so that holes in the end plates or central stays can allow for easy relocation to another hole or to be moved up or down in an elongated slot.

Although wings should be placed high up in the freestream, within the limits of series regulations, some sports racers place the wing very low at the rear of the car as it enhances the performance of the underfloor of the car. By creating a low static pressure underneath the wings' surface (which often has a large chord dimension when used for this function), a strong pulling force develops at the rear of the car and helps to suck air from underneath the car, increasing mass airflow and further reducing the static pressure (particularly at the diffuser throat) under the car, adding to the total downforce of the car.

4.4 Underfloor and diffusers

The underfloor of a race car is a vital area for producing downforce and can provide excellent efficiency results. Again, the key principle of the underfloor is to use **ground effects** to reduce the pressure underneath the car while ensuring we control its flow.

Ground effect: airflow underneath the body that is modified by the body running in close proximity to it.

All cars use ground effects, because they all interact with the air between them and the ground, however, some cars utilise this better than others. The key things to consider are ground clearance, underfloor shape and its roughness. Cars need to have a smooth underfloor in order to ensure minimum drag and maximum downforce.

A low ground clearance provides a more constricted area, so the lower the car the better as it helps to further accelerate the air under the car (and also lowers the CG). However, too low a ground clearance loses downforce due to viscous drag effects (although this should not be an issue with most as race series regulations set ground clearance at 40 mm). Some series specify a skid plate underneath the car, which allows for monitoring of ground clearance from when the car is in a dynamic state.

4.4.1 How do we get downforce?

We minimise air going under the car with the use of a splitter. The small amount of air flowing under the car will be in contact with the tarmac and so flowing only slowly. To get downforce, we use the rake of the car and the diffusers. A diffuser takes the air from the nearly flat underfloor and then gradually increases the volume of available space for the air to flow through by angling the diffuser upwards. This gradual increase in workable volume forces the pressure to drop, creating downforce. The pressure within it will be lower than the freestream, so some downforce is created, but the flat underfloor, with its large workable area, provides the core downforce. Thus a diffuser generates downforce through non-Bernoulli expansion.

The diffuser angle will have an effect on total downforce, but the key is to have the longest throat possible (constricted part of the venturi). This leaves you with the option of having a relatively steep and short diffuser angle, which is acceptable as long as flow remains attached.

Running chassis rake also allows for downforce to be optimised. You will find that running rake, where the front is lower than the rear, will provide not only greater overall downforce, but also allow for a greater balance of downforce to be placed towards the front of the car. What this does is create a mini venturi system under the car (adding to our larger picture of the whole underfloor acting as a throat), with the 'throat' being at the very front and the rest of the underfloor acting as a gradual diffuser in itself.

Front diffusers are also often fitted around the front floor area in order to provide additional front downforce near the outer edges, which allows the central part of the floor to work down through the whole length of the flat floor.

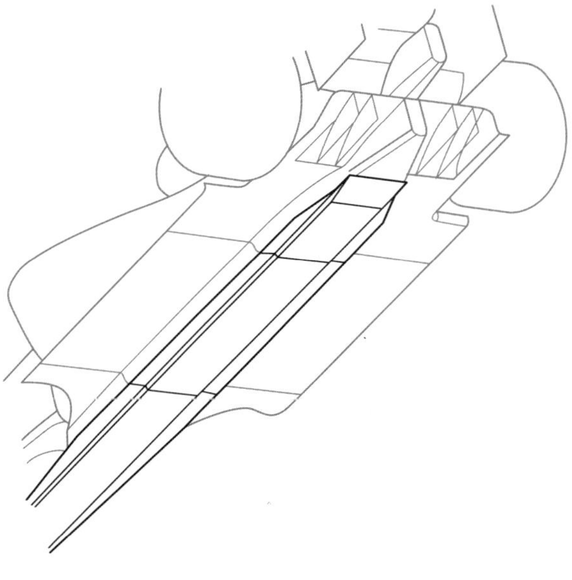

Figure 4.20 Flat underfloor and diffuser

Figure 4.21 Front diffuser

4.5 Additional parts

There are many other parts of a car that will help to create downforce. We must remember that any changes we make to the aerodynamics upstream of the car will have a direct effect on the rest of the car downstream, and so the aerodynamics must be designed as a full package to work together to produce a high efficiency with a stable aerodynamic balance.

If we take a look at two cars, a Le Mans Prototype (LMP) sports racer and a Formula 1 car, we can see a multitude of different aerodynamic parts. We are going to summarise what they do and how they work.

4.5.1 Splitter

This generally flat, horizontal-forward extension of the bodywork acts as a cutting device to separate the air going over and under the car. The splitter works by acting as an extension of the floor so that a low pressure can be created underneath it, while the upper surface acts as a shelf to harness the high-pressure air sitting around the front of the bodywork. Some cars have an adjustable splitter on sliders that can be either lengthened or shortened to act as a trimming device to reduce front downforce and change the balance. A smooth transition that leads directly from the splitter to the flat underfloor will ensure optimal performance, with some race cars separating the splitter from the front bodywork and instead making it part of the front floor.

Figure 4.22 An example of an LMP sports racer

Figure 4.23 An example of a Formula 1 car

Figure 4.24 Raised centre section splitter

A lot of LMP sports racers also use a raised centre section on their splitter, which allows for a greater volume of air to flow under the car and gain more downforce. It also ensures that the downforce levels stay consistent during ride height and pitch changes, because under dive or heave the front splitter could touch the floor and the underfloor become starved of air, causing the car to lose a significant proportion of downforce. Along with the raised centre section, some cars have tried to incorporate an underwing as part of their raised centre section, in order to create more downforce at the front as well as increase the total downforce.

4.5.2 Dive planes

These are commonly found on the outer corners of the front bodywork but are sometimes found on the rear edges. They are essentially curved plates that are angled upwards at the rear. They create downforce but also add a level of drag, dependent upon design and positioning. They are an easy-to-use trimming device and some cars run them in multiple tiers as well.

Figure 4.25 Dive planes

4.5.3 Arch louvers

These vents can be seen above the wheel arches and provide a simple but effective way of reducing lift to increase downforce. They reduce the high pressure that builds up in the wheel arch by pulling the air out, as the air pressure on top of the wheel arch is often lower than inside the wheel arch.

4.5.4 Wheel vents

These cut-outs behind the wheels (either in the side pods or at the back of the rear tub) work alongside the splitter, front diffuser and louvers to allow air to escape after being worked by the front diffuser. It is common to use this air to feed the coolers in the side pods.

4.5.5 Side skirts

These are horizonal extensions that protrude from the side pods, between the front and rear wheels. They are often used to help protect the low-pressure underfloor from the high pressure around the outside of the car body. The migration of air pressure could spoil the operation of the underfloor and affect downforce levels. Different shapes have also been tested to try and produce vortices along the side of the car to help seal the underfloor now that sealing side skirts are banned. Sealing side skirts linked the side pods of the car to the floor and sealed off the underfloor, which provided significantly higher amounts more downforce than modern underfloors.

Figure 4.26 Arch louvers and wheel vents

Additional parts **143**

Figure 4.27 Side skirt

4.5.6 Strakes

These are vertical plates that help to guide air. They are most commonly used in diffusers (both front and rear) to ensure that the flow does not become turbulent and to guide the flow away from the tyres or towards the wheel vents.

4.5.7 Barge boards/turning vanes/ vortex generators

These additional plates can be curved, angled, horizontal or vertical and placed in strategic places along the car. They work on the core principle of guiding the airflow in a certain

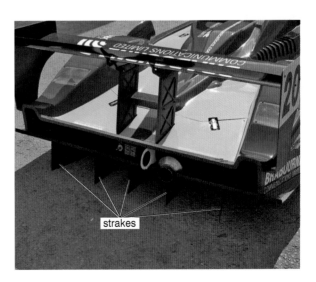

Figure 4.28 Diffuser strakes – vertical strakes can be seen on the underside of the diffuser

direction, whether towards or away from certain areas of the car. For example, they are commonly used in front of the tyres in a single-seater to reduce the drag as air is guided away from a high percentage of frontal area. Some may be used to force air into the cooling ducts, while others may target areas such as the rear wing or brake cooling ducts.

Aerodynamics, along with reliability, dictate the results of the high-end motorsport series. In most cases, they make the difference between coming first or second, and the team with the best aerodynamic designers and engineers is often the most successful.

Figure 4.29 Barge board

CHAPTER 5
Electrical

This chapter will outline:

- Sensing
- Wiring harnesses
- Batteries and components
- Driver interface
- Controller area network (CAN)

5.1 Sensing

Electrical sensors are used for a wide range of reasons in motorsport but mainly for the engine management system (EMS) and for data logging purposes.

We can sense (measure) anything that is measurable: temperature, pressure, position (linear or angular), speed, distance, acceleration, strain and many more. Sensor selection and application are key to create a system that works accurately and reliably.

We need to ask a variety of questions to determine what type and specification of sensor we require, for example:

- **What requires measuring?** Temperature, pressure, position, and so on.
- **What is the range of the expected measurement?** The values must be within the sensors' range (e.g. maximum/minimum of ±200 °C or ±3 G).
- **What environment will the sensor be exposed to?** A sensor can be sensitive to its environment and, if used in the wrong environment, it could produce an incorrect output or even fail. Local temperature and vibration often break the sensor, so its mounting needs to be considered, along with the likelihood of it being exposed to fluid, dirt and other contaminants.
- **What accuracy is required?** The level of accuracy depends on the requirements of what is being measured and whether the reading device (the logger) can find a benefit from it. Cost is a huge factor in sensor accuracy.
- **What is the budget for this sensor?** Sensors can be specified in different shapes, sizes and materials – the more money you have to spend, the better the sensor you can buy. Additional options include dual outputs, where there are two outputs for the same sensor which can be used for a variety of purposes.
- **Which size and weight is best?** This is often dependent upon the budget; the higher up the scale you go in motorsport performance, the more important the size and weight of these sensors become.
- **How fast is the sensor able to perform?** Later in this chapter you will find out how the logging unit can work at different sample rates. The reaction to change that the sensor can measure is key; for example, when trying to map an engine's fuelling map, it is important for the air intake sensor and lambda sensor to be able to measure and respond quickly.

All sensors should have an available specification sheet, which will include information such as the following:

- Transfer function – the relationship between physical input and the electrical output of the sensor, often shown in a graph.
- Sensitivity – the ratio of a small change in electrical signal to the small change in physical input (e.g. 500 mV/mm).
- Offset – the electrical output signal with zero input.
- Measurement range – as discussed above, anything outside of the range often results in inaccuracies.
- Tolerance – the largest expected error between the actual and ideal output signal. This range means that, for one input signal, there will be a voltage output range.
- Non-linearity – how far the signal can deviate from a linear output.
- Hysteresis – the variation of the output when the input value cycles up and down.
- Noise – all sensors produce varying amounts of noise, and this must be considered when viewing the output signal.
- Resolution – the smallest detectable input value change that the sensor can output.

Every sensor manufacturer supplies slightly different information on their specification sheets, but it should include information on the technical specification, mechanical data, electrical data, characteristics, installation notes and information on connectors and wires.

Data noise (as mentioned above) can vary from sensor to sensor and can be due to interference or an incorrectly calibrated sensor. This 'noise' will not be noticeable until it is analysed on the software package. Noisy data will contain more irregular and unexpected values, which can make the data potentially unusable as it is unreliable. To minimise the noise in the data, it can be filtered or processed to remove peaks and troughs that are too large or small for the data range that is expected. Most of the time, this is done automatically by the software. The only problems with this is that you are changing the data so may not be getting the full picture of what is happening, and some of the 'anomalies' may be actual values so the filtering will cover up the true results.

Another method of filtering data is to have a circuit built into the link between the noisy sensors and data logger that contains a series of transistors and capacitors to smooth the signal prior to reaching the data logger. In this way it will never receive the raw data, so removing the opportunity of the raw data being viewed before deciding whether or not to filter them.

It is also important to consider varying types of electronic control units (ECUs) or dataloggers and their input channels. Some have internal pull-up resistors and some can only accept raw voltage values. This will require some thought when selecting sensors and deciding whether to fit external pull-up resistors.

For more information on the operation and use of sensors, you can refer to *Hillier's Fundamentals of Automotive Electronics*.

A range of popular types of sensors are shown opposite and below.

5.1.1 Temperature sensors

These can be classified as contact and non-contact sensors. Contact sensors have to be in physical contact with the medium that needs measuring, such as oil, water and air temperature. Non-contact sensors are often used for measuring tyre temperatures and brake disc temperatures. Contact sensors used for temperature measurement are thermocouples, thermistors or resistive temperature devices (RTDs). Non-contact sensors are infrared (IR) sensors.

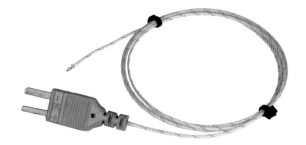

thermocouple

resistive temperature device (RTD)

Figure 5.1 Types of temperature sensor

infrared (IR) sensor

Thermocouples

These work by welding together two wires of different materials into a junction, called the measurement junction. On the other end of the signal wires is another junction, the reference junction. A change in temperature within the measurement junction generates a current in the wires proportional to the temperature change. Temperature at the measurement junction can then be determined from the type of thermocouple used (J, K, E or T type), the magnitude of the millivolt potential, and the temperature of the reference junction.

Thermocouples have a large temperature operating range and are reliable when exposed to vibration and shock because they are very simple in construction.

Thermistors

Thermistors change their electrical resistance in relation to their temperature. They have two metal oxides which are encapsulated in glass or epoxy.

Two terms are associated with thermistors: negative temperature coefficient (NTC), where the resistance decreases as temperature rises, and positive temperature coefficient (PTC), where the resistance increases with a rise in temperature.

Thermistors have a high sensitivity due to the large change in resistance but with a smaller measuring range than a thermocouple. The relationship between temperature and resistance is not linear. Thermistors are one of the most accurate types of temperature sensor.

RTDs

These work on the same principle as a thermistor in that a change in electrical resistance is used to measure temperature. The element used for sensing the temperature change consists of a wire coil or deposited film of pure metal, whose resistance has been documented at various temperatures. The most commonly used materials are platinum, nickel and copper.

RTDs have a similar measuring range as a thermocouple with the sensitivity of a thermistor, but their construction makes them fragile in harsh environments with high amounts of vibration.

IR sensors

Anything with a temperature above $0\,°C$ emits IR radiation proportional to its temperature. The emissivity (level of radiation emitted for a set temperature) of the body being measured must be known in order to determine how the sensor needs to operate. The IR sensor core converts the radiation energy into an electrical signal. Like a thermocouple, the change in temperature creates voltage.

Key benefits of IR sensors:

- Fast response times
- Ability to measure temperature on an otherwise inaccessible moving target
- High measurement range
- They do not lose heat energy when reading from an object

Disadvantages of IR sensors:

- Dirt and dust can cause inaccuracies with temperature measurement
- Only surface temperatures can be measured
- The emissivity of the measured body must be known in order to accurately measure the temperature

5.1.2 Pressure sensors

Pressure sensors are used to measure fluid pressures including oil, brake line (front and rear), coolant, fuel and air pressures for the manifold and for aerodynamic measurements (see pitot tube). Strain gauges are used in the form of a piezoresistive semiconductor. A diaphragm is used as the contact with the measured substance and the resistance value changes as pressure is applied to the diaphragm. The diaphragm's thickness determines the range of the sensor – the thinner the diaphragm, the lower the value of pressure that can be sensed; the thicker the diaphragm, the higher the value of pressure.

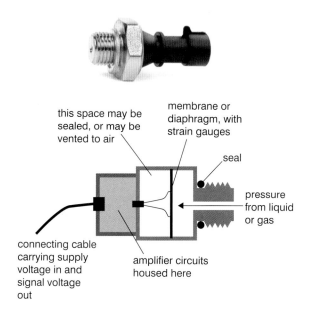

Figure 5.2 Pressure sensor

5.1.3 Displacement sensors

There are two categories for this type of sensor: linear and rotary.

Linear sensors tend to measure damper travel and wheel travel, steering angle and pedal positions. Rotary sensors can also measure pedal positions and steering angle, along with gear position on a sequential gearbox and throttle position from the throttle bodies/carburettor.

Linear and rotary potentiometers (pots) work on the same principle: as the sensor rotates or changes in length, it moves an internal slider (wiper) on a resistive element that is supplied with voltage. Thus, voltage proportional to the angle or length can be measured. These sensors are effectively variable resistors that react to a change in length or angle.

String pots operate in the same way as rotary pots but are activated by a retractable string, so can operate in areas where it would be unachievable with a conventional linear or rotary pot. Pulleys and guides can be used for this type of sensor to aid measurement.

linear pot

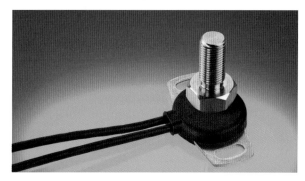

non-contact rotary sensor

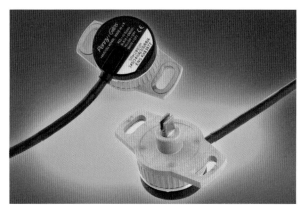

rotary sensor

Figure 5.3 Linear, rotary and string pots

magnetic ring sensor

string pot

The disadvantage of contact sensors is that they will eventually wear out. However, relatively new technology includes rotary contactless sensors, in the form of a magnetic ring that fits around a shaft, or a magnet that can be embedded into a bolt head or on to the end of a shaft for example. These sensors rely on a Hall effect chip to pick up a magnetic field. They have no moving parts so do not wear out and are also insensitive to dirt. The diametrically magnetised magnet ring can be used to measure steering angle and gear position, while the standard non-contact system can be used for similar purposes.

Linear variable differential transducers (LVDTs) and rotary variable differential transducers (RVDTs) are common in high-end motorsport set-ups.

An LVDT displacement transducer consists of three coils – one primary and two secondary coils. Current is transferred between the primary and secondary coils. Both LVDTs and RVDTs are controlled by an armature, which is a magnetic core.

When the sensor is in its middle position (e.g. 50 mm when total travel is 100 mm), the two secondary outputs are equal and, as they are connected in opposition, the output from the sensor is zero. When the magnetic core moves from its central position, the secondary coil it is moving towards will increase in voltage, while the secondary coil it is moving away from will reduce in voltage. This difference in voltage will produce an output from the sensor.

LVDTs have the ability to sense the position (or directional movement) of the magnetic core from the phase of the output, compared with the excitation phase.

The advantage of LVDTs and RVDTs is that there are no moving parts or electrical contact. This allows the component to have a long lifespan, excellent resolution and a very clean signal.

An ECU is required for these units in order to provide a source of signal conditioning. This allows you to control all elements in order to gain an output voltage, an output current, or serial data that is proportional to the actual position of the differential transducer. It drives the sensor with an alternating current (AC) power and can then convert the sensor's signal output from a low-level AC voltage to a high-level direct current (DC) voltage, which is more convenient to use. It also decodes the directional information of the sensor and can be programmed to determine the zero position.

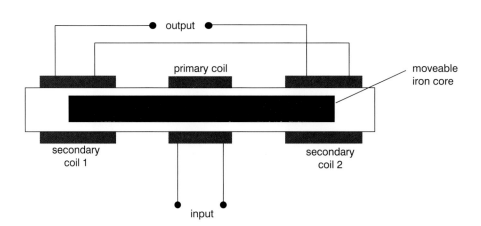

Figure 5.4 Linear variable differential transducer (LVDT)

5.1.4 Acceleration sensors

The key objective for this type of sensor is to measure longitudinal and lateral acceleration (and sometimes also vertical acceleration). It is the acceleration sensor that allows us to draw the G-plot mentioned in Chapter 3 Chassis and Chapter 4 Aerodynamics. Basically, a mass is housed centrally and secured by springs (usually made of silicon). When the device is accelerated, through vehicle braking, cornering or acceleration, it compresses one of the springs holding it in place; for example, when braking, the mass will move forward and compress one of the springs. The mass and the body holding it in place have conductive material on their faces and, when used together, they become an electrical capacitor.

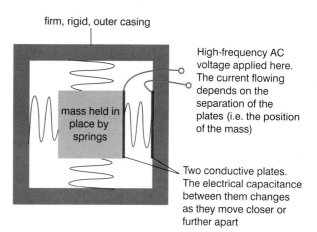

Figure 5.5 Acceleration sensor

A high-frequency AC voltage is applied to the plates; the current depends on the separation of the plates. The two conductive plates, therefore, change in electrical capacitance as they move closer or further apart. Capacitors allow AC to pass through them, but this depends on the capacitance of the capacitor. The capacitance increases if the electrically conductive areas are closer together. We can measure the compression of one of the springs by sensing the change in AC. For this reason, the compression of the spring is proportional to the acceleration.

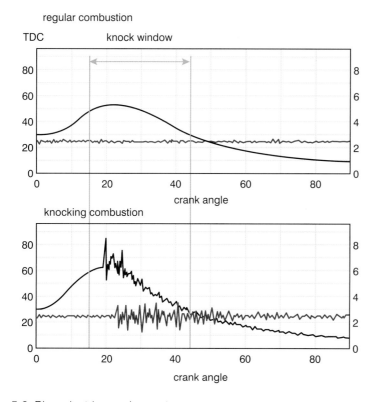

Figure 5.6 Piezoelectric accelerometer

These sensors are sensitive to tilt, so must always be placed flat in the car and as near to the centre of gravity as possible, allowing them to measure the stress the car is being put under as accurately as possible.

Piezoelectric accelerometers (see figure 5.6) are another type of acceleration sensor and are used to measure engine knock; they are effectively vibration sensors.

5.1.5 Speed sensors

Engine rpm and wheel speed are the most common parameters to be measured on a race car. Wheel speed sensors are normally put on the undriven wheels. If only one is used, it should be installed on the wheel with the greatest amount of load passing through it, other than brake force (conventionally, this would be the front left wheel on a clockwise circuit). This should mean that the wheel is constantly rotating with the speed of the car and so will be the most accurate. One speed sensor per wheel is, however, the optimal set-up to use.

Inductive sensors

An inductive sensor is a type of speed sensor. It has a soft-iron core surrounded by a winding. The sensor is positioned directly next to a toothed pulse ring that rotates with part of the driveline assembly. A very small air gap separates the two parts of the sensor assembly.

The soft-iron core of the sensor unit is connected to a permanent magnet which creates a magnetic field. The magnetic field reaches to the ferromagnetic pulse ring and is influenced by its position. When a tooth from the wheel is positioned opposite the sensor head, the magnetic field is amplified, whereas when a gap between two teeth is placed next to the sensor head, the magnetic field is reduced. As the toothed ring rotates, there is a constant change in the strength of the magnetic field. The magnetic flux changes occur in the transitions between tooth and gap and gap and tooth. As these changes occur, an AC voltage is created in the coil and the frequency that is produced can be used to calculate speed (see Faraday's law and Chapter 5 in *Hillier's Fundamentals of Automotive Electronics* for more information).

Figure 5.7 Speed sensor

This speed sensor produces its own voltage and sometimes has a shielded sheath over the wires to protect it from radio interference and to keep the signal consistent.

One output pulse per tooth is generated by the sensor. The amplitude of the pulse is determined by the air gap between the ring and the sensor, along with the rotational speed of the toothed ring, tooth shape and the materials used. Both the amplitude and frequency of the output signal increase with speed.

This means that a minimum rotational speed is needed in order for accurate measurements of wheel speeds to be assessed – at wheel speeds of around 2–3 mph, the sensor may not be accurate as very small voltages are produced. On the toothed ring itself, it is common to see a form of reference mark, which usually can be seen as a large tooth gap; this allows you to determine the position of the toothed ring. While this is not important for wheel speed, it is essential for camshaft and crankshaft position.

Any metal particles that are attracted to the sensor casing by the magnetic field must be cleaned off to avoid any chance of miscalculation. Maintaining the correct air gap (distance between sensor and toothed wheel) is also vital for the system to operate effectively.

Active wheel speed sensors

This Hall effect sensor has better durability, performance and lower speed accuracy than the inductive sensor. A toothed wheel has been used to trigger the sensor, although vehicles now tend to use a magnetic encoder instead. The encoders use north and south pole magnets embedded into a ring and operate on a magnetoresistive principle. Regardless of the type of trigger system, the outcome is a digital square wave and, as speed increases, its frequency increases but not its amplitude.

The operation of the system is energised by the ECU, which sends a small voltage to the sensor that gives it power. The sensor then has a feedback circuit, which returns the voltage (usually around 7 mA). As the magnetic encoder or toothed wheel rotates, the feedback circuit voltage is switched on or off.

The benefit of this sensor type is that it can sense rotational speed all the way down to 0 mph, whereas the inductive sensor type needs at least 2–3 mph in order to produce a reliable signal.

The sensitivity to change of the air gap is also lower, which is better suited to the harsh environment that rotational speed sensors are generally installed in, so there is less chance of error.

5.1.6 Strain gauges

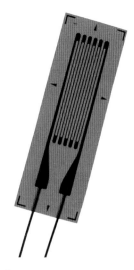

Figure 5.8 Strain gauge

Stress and strain is often an area that needs measuring in order to understand the forces that components are put under during dynamic motion, whether on a circuit or test rig. When strain occurs from various stresses, we can use a strain gauge to measure any mechanical motion of a part and turn it into an electrical signal. Strain gauges are often used on suspension components, drive shafts, steering linkages and aerodynamic parts. They can also be used as a way of cutting ignition when load is applied to the gear lever in order to provide clutch-less upshifts. As the metallic conductor placed on the component is deformed from its original shape, the electrical resistance will change.

Pressure sensors often use this technology to measure movement of the diaphragm.

Strain is determined by the ratio of the deformation of the original length compared to the original length, and can be expressed as:

$$\varepsilon = \frac{\Delta L}{L}$$

Key: L = Length   Δ = Delta (charge in)
      ε = Strain

A Wheatstone bridge circuit and amplifier is needed to measure relatively small strain levels and, therefore, small changes in resistance.

5.1.7 Pitot tubes

These sensors are used for aerodynamic measurements and are essentially a differential pressure sensor. You will often see them facing forward just in front of the cockpit of a Formula 1 car. This system can measure dynamic pressure by comparing static air pressure with total pressure (static and dynamic). The device contains two concentric tubes: the outer one measures static pressure on its top surface, while the inner tube measures the total pressure by facing straight ahead in the direction of flow velocity. These two tubes are then linked to a differential pressure sensor that gives out a voltage relative to the dynamic pressure.

Figure 5.9 Pitot tube

5.1.8 Laser distance sensors

These sensors are most commonly used for measuring ride height and ground clearance across the car. They operate by emitting a laser beam to the target area (if measuring ride height, this would be the floor), which is then reflected back up to the sensor unit where it is passed through a lens and into a photodiode. A change in ride height is determined by a change in angle of the returning beam received at a different part of the photodiode. A microcontroller will then calculate the distance to the target from the reflected beam's location on the receiver and will then output a voltage proportional to the target distance.

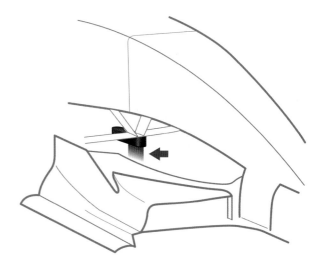

Figure 5.10 Laser distance sensor

5.1.9 Oxygen sensors

Also known as lambda sensors, these produce a voltage pulse when no oxygen is present in the exhaust gas; this occurs only when the air/fuel ratio is on the rich side of stoichiometric (14.7:1). A digital pulse signal is conveyed when no oxygen is present in the exhaust gas.

A voltage pulse from the sensor is used when the oxygen content is below a predetermined level. This pulse commands the fuel system (via the ECU) to reduce the fuel supply so that the fuel mixture is weakened, or it can just be used purely in a data logging system to log air/fuel ratio throughout an event. When used to alter the fuel supply, shortly after the alteration

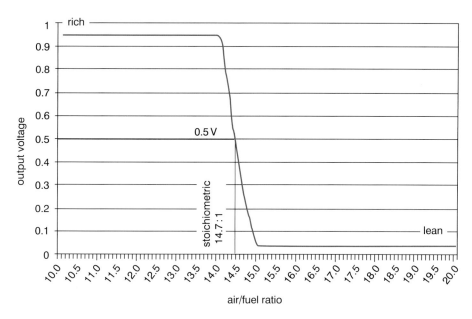

Figure 5.11 Lambda sensor

the exhaust gas oxygen sensor will detect that oxygen is present in the gas; this will cause its voltage output to fall and, as a result, the output from the sensor will change to zero. The ECU will detect that the signal has stopped and will then instruct the fuel system to enrich the mixture. In this way, the mixture is controlled by oscillating between the predetermined lean and rich limit; the ECU can then keep the air/fuel ratio within a range that is very near to optimum.

As modern sensor technology improves, we find that they are able to work to very small tolerances when switching the air/fuel ratio, which allows for exhaust emissions and engine performance to be optimised for its application. It is possible to have a limit of 0.05 from the required fuelling value for most sensors.

The rapid switching in voltage either side of lambda 1 gives a limit-control type of sensor cycle. In this case, the sensor can be considered to be a switch, which simply opens and closes to indicate rich and weak mixtures. When a digital output is used, the sensor will signal the two states showing 0 for the weak and 1 for the rich mixtures.

Electronic fuel injection allows the rapidly switching sensor to be used by the ECU to act upon the signals received so that fuel mixture can be altered without delay.

The sensor itself has a specific operating range and works inefficiently when below 300 °C. As a result, the ECU switches to an open-loop type of management for the fuelling system. This mode is used during cold starting, when the engine is idling, and if the control system detects a fault with the oxygen sensor circuit. Operating in open loop can be harmful to the lifespan of the catalytic converter (if fitted).

When accelerating and decelerating, the engine will generally operate in fuelling conditions outside of the oxygen sensor's range, requiring other sensors within the engine management to aid with the running of the engine.

Heated lambda sensor

Internally heating the lambda sensor gives more accurate control of the mixture when the exhaust temperature is below 300 °C. This occurs during cold starting and when the engine is operated at part-load for long periods. The heating feature also allows the sensor to be mounted further away from the engine. It is claimed that this cooler position gives this sensor type a life in excess of 100,000 km, which is an improvement; a sensor mounted close to an engine is exposed to temperatures in excess of its rated maximum (850 °C) during long periods of full-load operation.

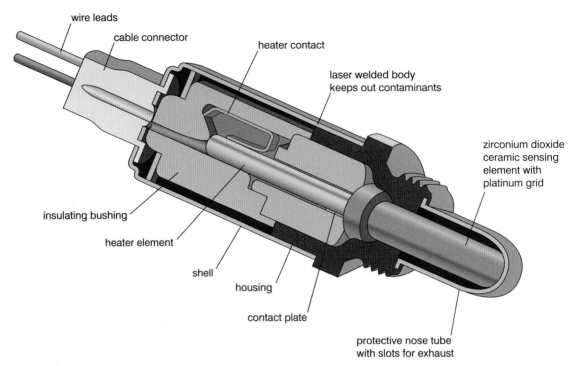

Figure 5.12 Oxygen sensor

The centrally positioned electric heating element, supplied with energy from an ECU, heats the inner surface of the active sensor ceramic. When cold starting, the heater brings the sensor up to its working temperature in about 30 seconds.

5.1.10 Global positioning system (GPS) sensors

GPS sensors allow us to determine positional and speed data. Global positioning satellite systems allow you to visualise the position of the car on the circuit within an accuracy of approximately 1 m. To gain the best results, the GPS unit should be placed so that it has a clear view and no obstruction overhead, and ideally a clear view around it as well. This is why GPS sensors are positioned on the roll bar of most sports racers and single-seaters. A GPS sensor is a more accurate way of measuring vehicle speed; wheel speed sensors can sometimes be inaccurate due to loadings through the tyre changing the effective radius of the tyre and so providing an inaccurate speed signal. GPS has become very important, especially to the low budget racer, as it enables lap timing, driver analysis, speed input, and so on.

Figure 5.13 GPS sensor

5.2 Wiring harnesses

The wiring harness (or loom) of a race car should be low maintenance and reliable so that a race team can focus on the areas of the car that require attention at a race event, such as chassis set-up and general maintenance and preparation.

Wiring harnesses must be robust and flexible and, when constructed, should be laid up so as to facilitate flexibility and to minimise the overall size. They need to survive in harsh environments, which include high temperatures, vibrations, rain, dirt and exposure to other fluids. Due to the amount of harness required, they must also be light as most modern race cars have an extensive data logging system, driver interface/dashboard system and of course the engine management system along with ancillaries, such as transponders, intercoms and rain lights. Whether a complete bespoke harness is made, or parts of the original harness are reused, it is important to simplify the design as much as possible so that anything that is not needed is removed, and anything that does not look robust should be replaced.

Multi-strand wires are preferred over single-core wires due to their flexibility, although they do not have such a high current-carrying capacity. Another factor in selecting the wire size is the resistance to current and hence the drop in potential difference across it. The current rating for each cable needs to be considered based on the equipment that it is going to be connected to.

When connecting wires together, ensure that the connection has a low electrical resistance, good mechanical strength and protection from the environment.

Crimping terminals and high-quality connectors, such as Sure Seal, Deutsch and MilSpec multi-plugs, are robust, waterproof and lightweight. The selection of any connector should be governed by the importance of a breakdown; so anything that will potentially stop the car (such as the engine) should be given budget priority over data logging systems, so a better-quality connector should be used. Any connector that is regularly being connected or disconnected should be of a higher quality too. The correct tools for crimping and inserting or removing each individual terminal and pin should be used to ensure reliability.

Figure 5.14 Wiring harness elements and also shown as connected to a power panel

The harness should be made in subsections so that sections can be easily maintained, replaced and inspected. Using subsections will also allow for ease of overall maintenance to the car. If an engine needs to be replaced, having a separate engine harness that can be disconnected from the car with one multi-plug would make the job much simpler.

Using connectors is often preferred to soldering as it provides a way of crimping the wire, but also supports the outer sheath. Solder on the other hand provides a good electrical connection but can also cause a weak point either side of the solder which could fatigue and break due to vibration and bending. The outer sheath can also become damaged from the heat of the soldering process.

The wiring harness will need to be routed away from high heat sources (e.g. exhaust) and moving parts (e.g. pulleys, belts and drive shafts) in order to avoid damage and minimise interference. Cable tie mounts and P-clips are commonly used to safely secure the harness to chassis tubing, panels and bodywork.

To encase the multiple wires, a high-quality heat shrink tubing is most commonly used, such as Raychem DR25, which is flexible and abrasion resistant. It has high-temperature fluid resistance, and a long-term heat resistance and can shrink down to 50 per cent of its supplied diameter when heated. It is available in straight lengths or angles, reducers, and so on. On the ends of any DR25 you should use some RT125 adhesive, which is a flexible, two-part adhesive that cures at room temperature and is used for general-purpose wire and harnessing applications. This shields the wires from fluids or dirt. You can also use adhesive-lined heat shrink, but this ages rapidly.

Additional protection can be used in the form of glass fibre or Kevlar heat protection sheathing if cabling does have to run close to a heat source. Braided, convoluted and spiral-type outer protection can also be used.

Labelling the wires is vitally important in order to help trace any faults or aid with interchanging components. For this, the best choice is heat-shrink labels – clear tubes of heat shrink and a simple coloured strip of label (that can be written on with a label maker), which are slid over the top of the normal heat shrink. Cable identification within a wiring harness is often determined through a simple colour-coding process. Red wires are live, black wires are battery negative and white is a signal wire. For the white signal wires, you can use coloured identification rings to link them to a key chart.

Remember the Motor Sports Association (MSA) requirement is that battery cables must be identified with red for positive and yellow markings for earth; a small strip of coloured heat shrink over the top will do the job. The terminals also need to be covered and the battery held firmly in place, otherwise you are likely to be pulled up by the scrutineers.

5.3 Batteries and components

The battery and the other electrical ancillaries, such as relays, voltage rectifiers, data logging control units and ECUs, should be mounted securely and ideally away from where damage could occur from a crash, water, other fluids and constant dirt and debris sources. Most should have some type of damping in their mounts in the form of rubber washers so that the internal circuitry does not fail.

5.3.1 Batteries

Batteries are compulsory in a race car in order to keep all of the electrics supplied with power (spark plugs, etc.), although they do weigh quite a lot in racing terms. However, lithium ion technology is advancing and these types of battery are starting to become available for motorsport use (although at a high cost) and tend to be significantly lighter than a conventional battery of the same specification.

Most race batteries are sealed and constructed using an absorbent glass matt (AGM), which separates the positive and negative plates. The electrolyte (battery acid) is absorbed by the AGM, meaning that the battery can be mounted at any angle and, if it splits, will not leak any acid. They do not like vibration and care should be taken to reduce this, especially in single-seaters.

Batteries' key specifications are ampere hour rating, cold crank ampere rating and pulse current rating.

Ampere hour (A/h) rating

This defines the actual capacity of a battery. A battery that is rated at 100 A/h at the 10 h rate of discharge is capable of delivering 10 A for 10 h before the terminal voltage drops to a standard value such as 1.67 V per cell.

If you then multiply the 1.67 V by the six cells that make up a 12 V battery, you get the end voltage, which in this case is 10.02 V. At this point, the battery is considered discharged. On a similar note, a 50 A/h battery would be able to supply a 5 A load for 10 h.

Cold crank ampere (CCA) rating

This is the number of amperes that the battery can supply for 30 s at a temperature of 0 °F (−17.8 °C), until the terminal voltage drops below 1.2 V per cell or 7.2 V for a 12 V battery.

Pulse current rating

This is the number of amperes that a battery can deliver for 5 s until the terminal voltage drops below 1.2 V per cell or 7.2 V for a 12 V battery.

An Anderson plug can be used to provide a slave facility to help start the car when hot. This allows for direct connection from the on-board battery to a fully charged battery that is carried on a trolley by a team member. You can also use a twin battery set-up to give 24 V starting for high compression engines.

Some cars can get away without using an alternator if there is not a high demand of current from electrical systems. Cars with carburettors, few or no lights and a simple dash and data logger set-up do not produce a high demand for current, and so should be able to use a high-capacity battery that is fully charged for each sprint race (e.g. up to 30 minutes) and not run an alternator, which adds weight and draws some power from the engine.

You will also notice that a Formula 1 car does not have a starter motor fitted and relies on a highly powered 24 V external starter motor system to turn over the high compression and small tolerance engine. The car itself will then use an alternator and small battery to power itself during the race.

5.3.2 Rain lights and transponders

Some compulsory electrical items to be aware of for circuit racing are the rain light and timing transponder. The MSA Blue Book (*Competitors' and Officials' Yearbook*), which sets out all of the rules and regulations for British motorsport, determines the standards and positioning of these two components.

At the time of writing, the rules for rain lights are:

> A rearward facing red warning light of a minimum of 21 watts, with surface area minimum 20 cm^2, maximum 40 cm^2, or of 21 watts with a surface area minimum of 50 cm^2 and with lens and reflector to EU Standards, must be located within 10 cm of the centre line of the vehicle and be clearly visible from the rear. Vehicles fitted with full width bodywork may alternatively use two lights equally located about the vehicle centre line. An alternative light unit of equal or enhanced constant luminosity or LED lights that are either homologated by the FIA or comply with relevant EU Regulations may be used. The warning light must be switched on when visibility conditions are reduced, or as detailed within championship and/or event regulations, or when so instructed by the Clerk of the Course.
>
> *2012 Competitors' and Officials' Yearbook,* Motor Sports Association, UK

Figure 5.15 Rain lights

The guidance for the transponder is that an AMB TransX 260 transponder should be used in circuit racing, either directly powered by the battery (which is the most convenient option) or with a rechargeable system. The transponder is used to log official lap times and each transponder has a unique number that must be registered to the car prior to the race.

The transponder should ideally be positioned as close to the surface of the track as possible and placed vertically. The transponder should also have a clear view to the track surface directly below it, so should not have any form of metal or composite between it and the track – as this can sometimes interfere with the signal transmission.

The directly powered units have the benefit of moulded mounting lugs so that the transponder can be bolted to the chassis of the car. When installing, the directly powered unit needs to be connected to a fused 12V supply from the car's battery. When the battery isolator is then turned on, the unit will be permanently active and emit a green light to show that it is powered. The unit uses normal UK household wiring colourings: brown is positive and blue is negative. It will also draw no more than 30mA when live. (Information taken from Timing Solutions Limited.)

5.3.3 Battery isolators (or master switches)

Battery isolators are also required by the MSA and must isolate all electrical systems when turned off, with the exception of the fire extinguisher system, if electronic. The regulations at the time of writing are as follows:

> The circuit breaker, when operated, must isolate all electrical circuits with the exception of those that operate fire extinguishers. The triggering system location must be identified by a Red Spark on a White-edged Blue triangle (12cm base), and the 'On' and 'Off' positions clearly marked.
>
> *2012 Competitors' and Officials' Yearbook*, Motor Sports Association, UK

Battery isolators can be either electronically operated or mechanically operated by hand.

When the electrical system is activated/turned on, the isolator connects the negative terminal of the car battery to the chassis, which provides power to all of the electrical systems on the car. When deactivated/turned off, the isolator will disconnect the negative terminal of the battery from the chassis as well as the alternator feed from the positive battery terminal. It usually has a fail-safe system whereby, if any of the STOP switches become faulty, the system turns off and kills power to the car.

With a manually activated system, the positive feed from the battery goes to one of the large isolator terminals; on the other side is the feed to the electrical systems and alternator. The secondary circuits, often stamped as W (1) and Z (2), work in opposing ways. The W circuit, when switched on by the master key, is in open circuit, but when switched off, the circuit is closed through a resistor to earth. This allows for the other terminal to be linked to the alternator and to protect the alternator diode pack from blowing by dumping its current through the resistor. The Z circuit, when switched on, is in closed circuit, and is in open circuit when the master key is in the 'Off' position. This circuit allows the ignition switch to provide power to the ignition coils.

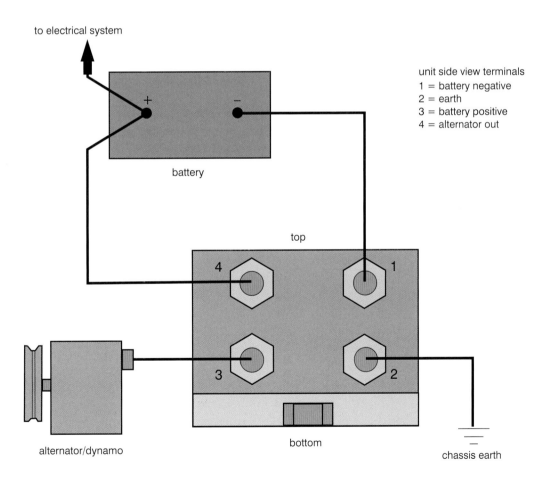

Figure 5.16 *Electronically operated isolator*

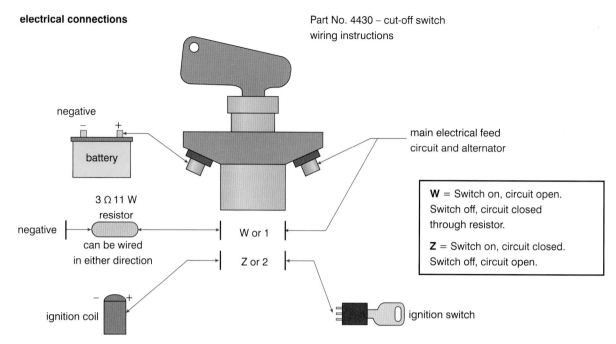

Figure 5.17 Mechanically operated isolator

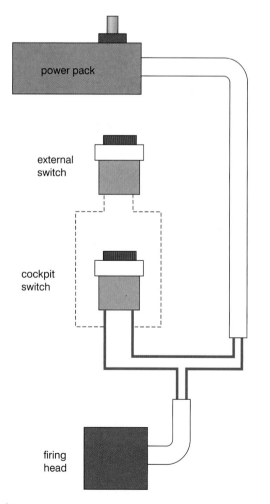

Figure 5.18 Electronic fire extinguisher

5.3.4 Electronic fire extinguishers

Electronic fire extinguishers are often used in race cars, and these must NOT be controlled by the battery isolation switch but should be directly wired to the battery. The only electric part of the fire extinguisher system is the actuation system. The push-button activation (often situated inside the car as well as outside) triggers an actuator that opens the valve of the pressurised fire extinguisher chamber and allows the unit to discharge into the engine bay and cockpit. The systems often have a control box with a three-way toggle switch to allow three modes: activated, deactivated and test mode.

5.4 Driver interface

The electrical interface for the driver will include the switchgear and dash display unit. It is important for this set-up to be simple to operate and in reach of the driver without distracting them from their key focus – driving fast and racing!

Switches must be well laid out, easy to operate and clearly labelled so that no mistakes are made

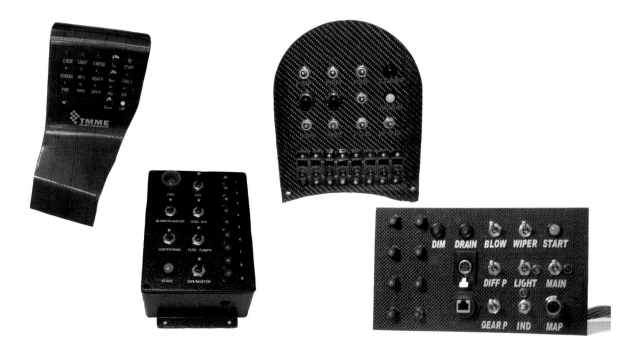

Figure 5.19 Some control panel and box layout examples

during the pressure of a race. Common switches to include in a panel would be the ignition, fuel pump and start button cluster. Rain light, driving lights (if fitted) along with the items already discussed in this chapter should also be included. In a saloon car, you could add heated screen elements, wipers, screenwash, horn, intercoms, and so on. Switches come in a variety of options, such as toggle switches and buttons. Most switches, especially cheaper alternatives, do not like vibration or high current. Relays should be used for switching high-current components. You will also often find some switches and buttons mounted to the steering wheel, to allow for quick and convenient switching or control of equipment on the car. They are often connected to the dashboard via a coiled flexible wire that allows the wheel to be rotated without damaging the wiring loom.

The switches themselves must be robust and fitted to a rigid switch plate so that the driver does not have to contend with a switch breaking if they are heavy handed with it. An alternative to using fuses in the main circuits that run the car are circuit breakers, which pop out when the current levels are exceeded; you can then push them back in to close the circuit and try to carry on.

The dash display is a vital area of the car to customise to the driver's requirements and can give the driver a great benefit if it can display what they need it to.

A lot of sports racers and single-seaters do not have a great deal of room for dash display units, but systems such as the Farringdon SWIS10 provide a great deal of information in a simple and easy-to-read layout. The sectioned windows allow you to quickly see what piece of data you are looking at and, with a minimum number of control buttons, the system is relatively uncluttered. The dash ideally needs to display key engine parameters, such as engine oil temperature and pressure, along with water temperature and perhaps fuel pressure. Rpm and shift lights, wheel speed, elapsed time, fastest lap and lap delta (which are very useful in qualifying) should also be included. Also, the number of laps completed, along with gear position, can be helpful. Some systems have multiple screen views, which allow the driver to flick through data screen by screen with the use of buttons that should be mounted to the steering wheel for ease of operation.

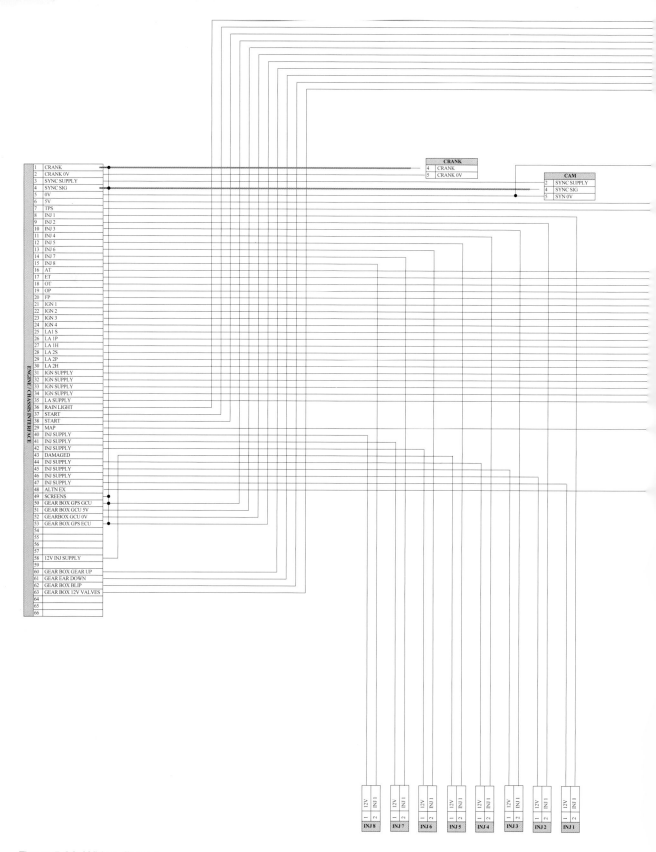

Figure 5.20 Wiring diagram

Driver interface

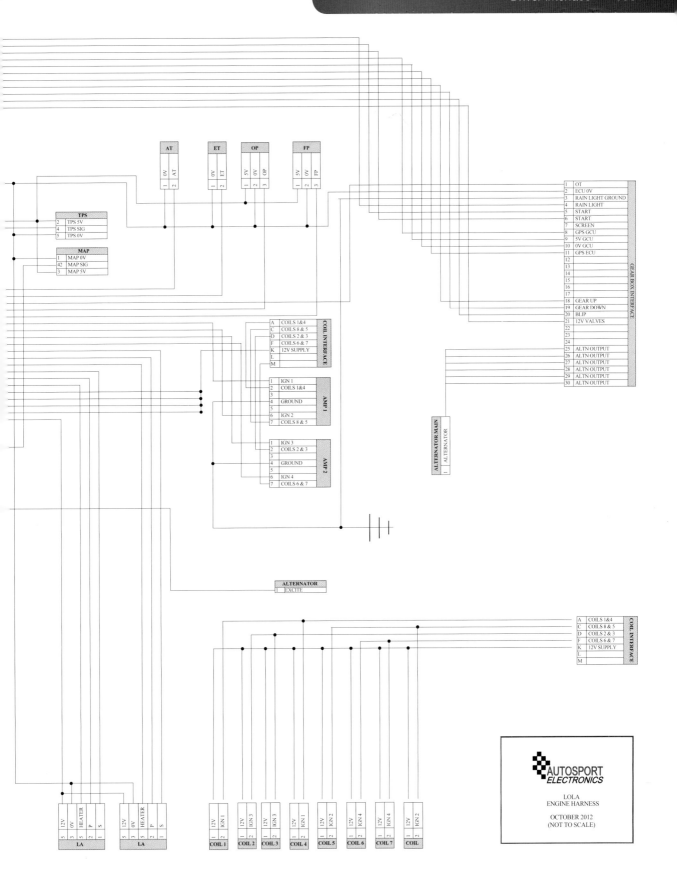

Figure 5.21 Farringdon SWIS10 steering wheel-mounted display

Another technology that is becoming common in motorsport is power delivery/management modules (PDMs or PMMs). These control boxes distribute power from the battery into the required channels. They replace relays, circuit breakers and fuses, while also simplifying the wiring harness, reducing weight and providing diagnostic capabilities. The unit has the ability to be programmed through a laptop to provide each channel with programmable fast and slow blow-fuse characteristics. The unit has the ability to reset any fuses and provide reports on current usage. The system can also provide the functional status of all channels to a control panel. The inputs (e.g. switches) and outputs to power the systems (which could include any of those discussed previously) are all fully programmable. The unit can have an engine-start function to allow full power to be directed to the engine-start channel and also a low-battery feature where you can programme the unit so that it prioritises power delivery to the systems that will keep the car going.

A controller area network (CAN) bus system links the PDM to a control panel on the dashboard, which is USB-programmed as to what the switches do. There is a light-emitting diode (LED) warning system for each switch, to warn the driver of any error detection.

5.5 Controller area network (CAN)

This technology has become key to the way electrical systems communicate with each other. CAN reduces physical wiring and enables a vast variety of communication between the ECU, data logger, PDM and sensors. See *Hillier's Fundamentals of Automotive Electronics* for more information on this topic.

CHAPTER 6
Data logging

This chapter will outline:

- Data logging systems
- Telemetry
- What to monitor – driver/chassis/engine
- How to analyse the data

This chapter discusses generic data logging systems, which use a variety of sensors and an electronic data logging unit. However, there are many less sophisticated ways to log data: away from the car, we can use a stopwatch or official lap timing; on the car, we can use a standalone video camera system or brake disc temperature paint, temperature stickers and many other indicators. Never forget that the most important sensor is always the driver, who needs to be able to feel and understand how the car is handling and how their driving style can affect the car. They must be sensitive to set-up changes, so working closely with the driver will make it a lot easier to develop the car.

6.1 Data logging systems

Data logging (or acquisition) gathers data from a variety of sensors and records the information for analysis at a later time. It can be an excellent tool for many reasons. It provides a method of recording parameters that are vital to the reliability and performance of the vehicle and its systems, and it can allow you to analyse the driver's performance individually or against a number of other drivers.

Data logging can be very expensive and is often underused in club racing, even when systems are installed in a car. By opting to potentially spend thousands of pounds on a data logging system with all of the associated sensors and wiring that adds a few kilograms to the car's overall weight, you need to ensure that you get the best from the system.

To be able to understand a data logger's results so that you can optimise braking performance, damper operation and aerodynamics, you must be able to understand the dynamics of the vehicle and its systems. This helps you to identify any potential issues and also related driving techniques so that you can advise the driver on how to improve lap times through throttle, steering and brake application techniques, along with shift points and other elements.

Most data logging systems not only provide a software package that allows the data to be viewed and manipulated in a range of methods, but also link in with a dash display system so that key data can be shown to the driver.

A data logging system will usually consist of the following:

- Logging unit – this is where the data from the sensors is converted and stored, along with the settings for the logging system. Data is either written to a removable card, such as an SD card, or held in an internal memory that can only be removed via a data cable. Some logging units are built into the dash display unit, such as the AIM and MoTeC systems, while others have a separate logging unit with a link cable into the back of the steering wheel dash display, such as the Bosch and Farringdon units.
- Sensors – a range of sensors are needed to log engine, chassis and driver-based parameters. The more sensors you have, the more it will cost. An engine will often have a range of sensors already fitted for the engine management system (EMS) that can be used. These are often linked to the logger unit via a controller area network (CAN) bus link.
- Lap beacon – this uses an infrared beam transmitter placed trackside and a receiver fitted to the car that points in the direction of the transmitter as it passes the pit wall. Some systems need their own transmitter, while others can rely on the circuit's own system. (Remember to switch the receiver over when you visit a circuit with the pit lane on the other side, for example, going from Donington Park to Brands Hatch.) To avoid inaccuracies, it helps to position your beacon away from others and set the 'lap blank' so that it cannot pick up another beacon within a set time period. Note that direct sunlight into the beacon can sometimes make the system inactive.
- Global positioning system (GPS) sensor – this sensor has the ability to determine speed and lap times (with the use of software), and draw an accurate map of the route taken by the car on a lap or rally stage.

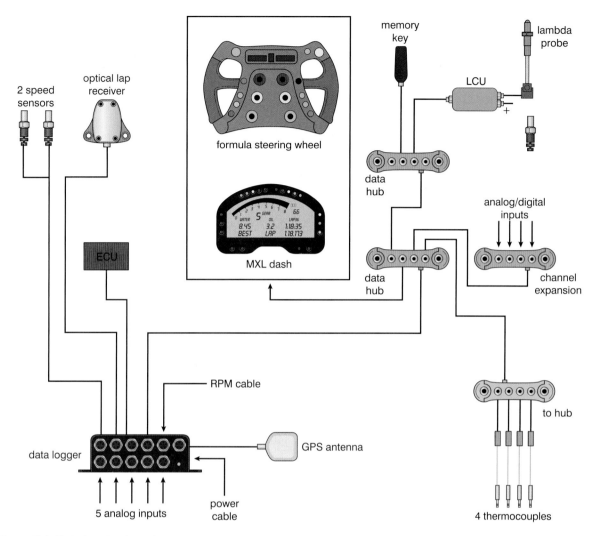

Figure 6.1 Data logging layout

- Software – this should allow you to view, compare and manipulate data in a variety of methods in order to create visual data to aid interpretation.
- Hardware – a laptop, computer or tablet capable of processing the data and running the software program. Always back up data to an external storage device (e.g. memory stick, CD-ROM, etc.).
- Camera recording system – this is a great add-on that can allow you to overlay data on to a race video, which is a great visual aid and can help to explain any anomalies in data.
- Expandable channel unit – this allows you to include more channels once the maximum number has been reached from the basic logger unit.

When determining what system you should purchase and how it should be set up, consider the following characteristics.

Logging/sample rates

This is how often the logger will store a piece of data. This is measured in hertz (Hz), which is rated in counts per second. Most systems will allow you to customise the logging rate for each sensor, as some sensors will require a higher sample rate than others. A low rate would be used for fluid temperature, which would not change its output a great deal over the course of a race, and we could assume a rate of 2 Hz, while a high rate of up to 200 Hz would be needed for a linear potentiometer measuring suspension travel as its output would be constantly changing at a very fast rate in both directions. Obviously, the higher the logging rate we use, and the more sensors there are, the more memory it will take up in the logger. The higher sample rate required for some sensors is used to provide more accurate data – as the sample rate is higher, more data is then taken and stored, allowing for a more accurate and in-depth analysis of the data showing up any anomalies more clearly.

Memory

Data loggers have either an internal or external memory, as discussed earlier. The memory capacity required will depend upon the size of the data logging system, in terms of the amount of sensors and the logging rates used. We can use the following calculation to work out the approximate memory capacity requirement:

> Bytes needed =
> 2 × No. of channels × Logging rate × Time (s)

If we assume we have a 20-channel logger with all sensors reading at 100 Hz for a 30-minute period, we can calculate:

> 2 × 20 × 100 × (30 × 60) = 7,200,000 b ≈ 7.2 Mb

Note that some video systems log straight into the logging unit and these use sufficient amounts of memory. Some systems can compress data, allowing data to be stored over a longer time period. For example, if a particular channel, say water temperature, did not change for a five-minute period, instead of storing the same temperature reading over and over again, the unit will produce a piece of code that states the temperature and the number of readings.

Channels

These are the amount of inputs that a data logger can read. The more channels the logger has, the more sensors you can use. Some systems may only be able to log eight channels, for example, but may have an optional expansion pack that increases the number of channels that can be logged.

Of course, the more channels, the more memory and the higher the sample rate that the logger can read, the more expensive it will tend to be. A typical club-level logger will accommodate 4–16 channels, while a Formula 1 car will have around 250 channels available.

Analogue to digital converters (ADCs) are used to convert the analogue signals received from the sensors into digital signals (0 V or 5 V) for the data logger to store. Not every channel suits this, such as damper plots where the data is over such a varied range.

A data logger can also be a good way of quickly diagnosing possible faults or issues with the car. With a view of the dashboard or by reviewing the data logging in the analysis software, you can easily view fuel pressure, battery voltage and other reliability-based systems.

6.2 Telemetry

Telemetry is an advance on data logging and allows for real-time viewing of the data in the pit lane and in racing, such as Formula 1. This data is then transmitted back to the factory base for a team of analysts to help determine set-up, strategy, and so on.

Engine parameters, tyre pressures and temperatures can all be analysed in real time and these can then be adjusted on the next pit stop or relayed back for the driver to change settings in the cockpit or driving style.

Two-way telemetry was used in Formula 1 until it was banned in 2003. This allowed engineers to change the engine's maps, alter sensor calibration and much more in order to improve performance in different conditions, or fix the car if it had a running fault while still lapping around the circuit.

Telemetry can be relayed back to the pit lane via a beacon, so every time the car passes the beacon data is transferred, or in real time, where data is continuously transferred as the car laps around the circuit.

Telemetry is also used in most basic data logging systems in the form of lap delta performance analysis. This system is present on the Farringdon system discussed in Chapter 5 Electrical. It calculates how far the car has travelled on a given lap and in what time, and compares this to your current fastest lap, thus giving a difference in lap time at that stage of the lap.

6.3 What to monitor – driver/chassis/engine

As a data engineer, you will be required to look at three systems: the driver, chassis and the engine. Although separate, they often overlap.

Driver performance sensors can include rpm, throttle angle, brake pressure, steering angle and acceleration sensors (longitudinal, lateral and sometimes vertical).

Chassis-related sensors can include steering angle, damper potentiometers, acceleration sensors, load cells and strain gauges, tyre temperatures, pressures and ride heights.

Engine-related sensors can include water temperature, oil temperature, oil pressure, fuel pressure, air/fuel ratio, rpm, throttle position and air temperature.

Systems such as the Race Technology DL1 has a basic package that already includes built-in accelerometers (longitudinal and lateral) along with a GPS sensor, which will allow you to add speed, track maps and therefore lap and sector times without needing additional sensors. You can also use an rpm trace from a coil pack, which costs next to nothing. This system then also allows you to have eight analogue sensors, four wheel-speed sensors and an electronic control unit (ECU) connection.

An entry-level set-up ideally needs to measure speed, rpm, lateral and longitudinal acceleration, throttle position and steering angle, in order to analyse basic performance of the chassis and driver, although some entry level systems do not have steering or throttle sensors due to their high cost. On top of this, you would want to measure oil and water temperature, oil and fuel pressure (to ensure engine operation is within certain parameters) and lambda/exhaust gas temperatures perhaps. Knowing brake pressure in each line (front and rear) is also advantageous and can allow bias to be calculated using a maths channel.

A mid-range system would include damper position for all four corners. A high-level set-up would include ride height, air speed, tyre temperatures and pressures, force acting on push/pull rods or dampers and yaw angle.

6.4 How to analyse the data

To analyse the data, we need software that:

- provides reports of channels that can show minimum, maximum and average values
- has the ability to display multiple channel data on the same or separate charts
- measures lap and sector times
- has maths channels
- can overlay laps or sectors over one another
- has a time slip function to show losses and gains across a whole lap or sector
- produces XY charts, G-plots and histograms.

The software needs to be easy to use and easy to customise to suit the user and their requirements.

All software is operated in a different way and using the software is the best way to become familiar with what the data looks like, how to best analyse each channel and what works best for you in terms of ease of analysis and the range of data available to you. You can download MoTeC Data Analysis software for free from www.motec.com/i2/i2downloads with sample data and configurations that allow you to familiarise yourself with the system. Some other suppliers also offer a free download of their software with sample data.

Track maps are most commonly created via a GPS sensor, but some systems rely on the lateral g-force and speed sensors to create an interpretation of the map. Track maps are a great visual aid as they allow the data to be more understandable than just being in a plain

172 Data logging

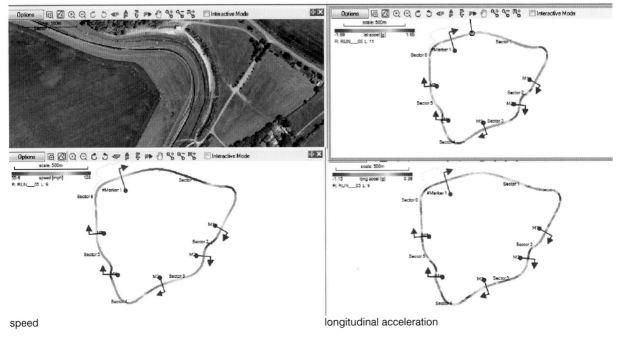

Figure 6.2 Track maps

graph form. They can also have a colour plot attached to them to show a data trace and its changing values over the lap, and with the use of a GPS sensor, show various lines taken with each lap/sector. You can also use this function to help the driver become reacquainted with the circuit by plotting shift points and braking points or to show the driver where errors were made throughout a session or lap so that they can improve.

Time slip is an excellent tool and shows time lost or gained against another lap/sector as a percentage, or in seconds. It is probably the easiest way to quickly target where time was lost or gained against the target lap. Any rise in the trace against the reference lap line shows that you are losing time to the reference lap, while the trace falling will show a gain in time.

Sector times can be used to break each lap time down into smaller chunks, allowing us to see where time was lost or gained on a specific lap. With the use of sector times, we can bolt all of our fastest sectors together to gain a theoretical best lap that the driver and car combination could be capable of – the ideal scenario being for the driver to lap during the whole race at the theoretical best lap pace. Obviously, this is not always possible in a race where other cars on the circuit can affect lap time, along with changing conditions and dynamic changes to the car, such as tyre temperatures and pressures along with any form of tyre wear, as discussed in Chapter 3 Chassis. It may also not be possible to perfect every corner, particularly those that flow into one another. For example, exiting the Brooklands corner at Silverstone brings you out on the wrong side of the circuit, depriving you of the best line for the Luffield corner. One corner jeopardises the line of the next corner due to their close proximity. In this example, the two fastest sectors for these corners could not be bolted together to give a theoretical best, as the racing lines would not integrate together. It is more accurate to use a smaller number of sectors, often known as splits, to use the best groups of sector times to provide the theoretical best. From this, you can split the track in a way that combines the corners that provide any kind of compromise to the next – two difficult corners are taken as a single corner in terms of time chunks, providing a more realistic outcome.

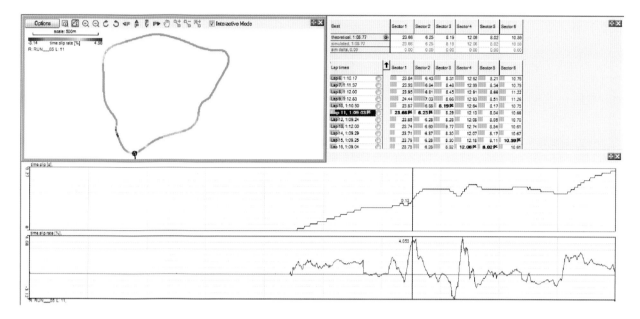

Figure 6.3 Time slip chart

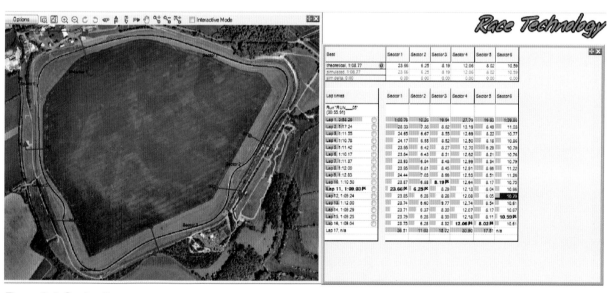

Figure 6.4 Sector times

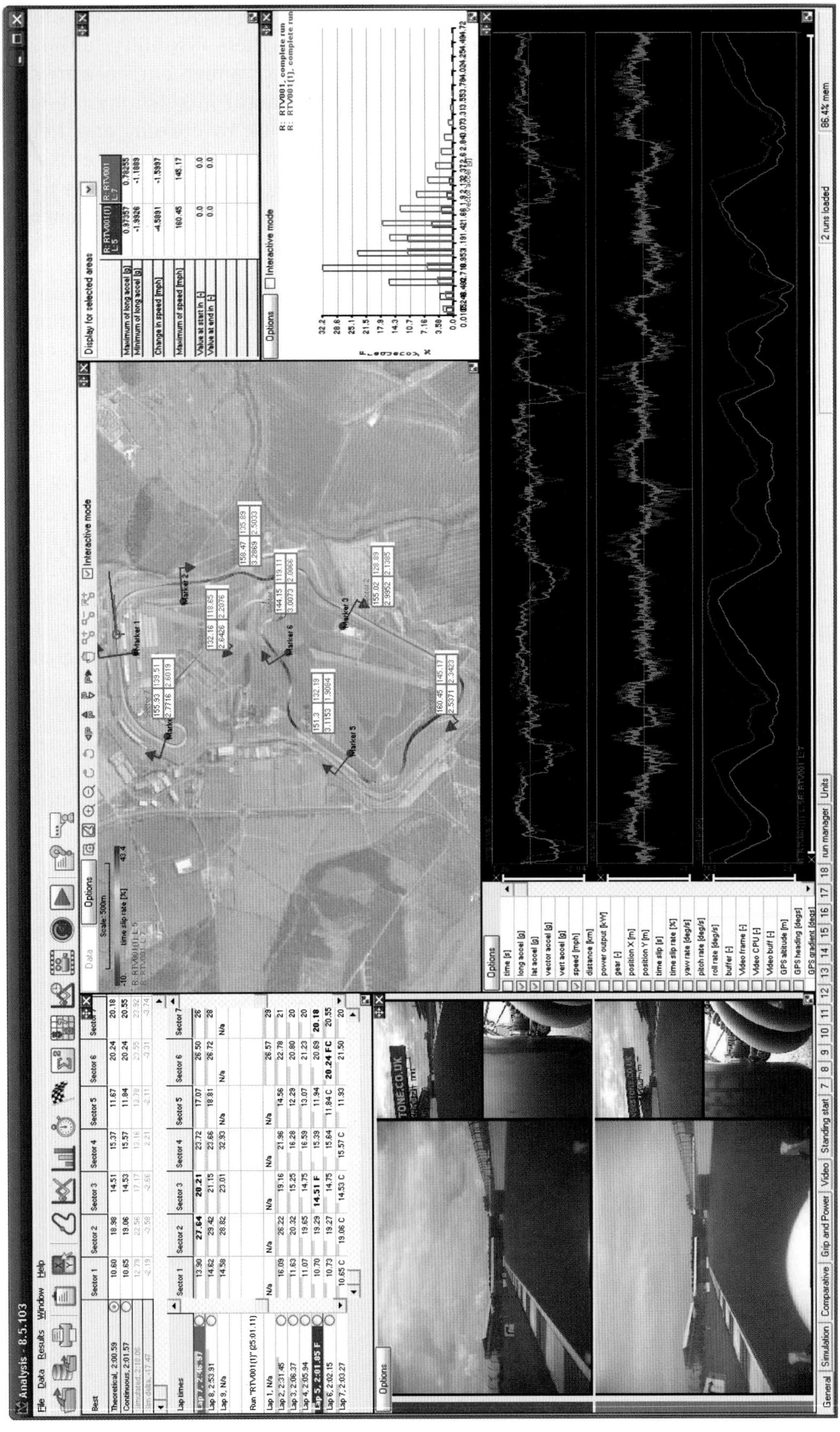

Figure 6.5 Analysis software can allow you to monitor various selected aspects simultaneously

Consistency of these laps, splits or sectors is key so you need to determine whether it is the car or the driver that is inconsistent. Overlaying inconsistent sectors of the lap via the speed trace will usually provide you with a quick view of where the inconsistencies are contained within that sector.

Channel reports can be used to show maximum, minimum and average values, which are particularly important for assessing the health of the engine with parameters such as rpm, oil and water temperature, oil pressure, air/fuel ratio and fuel pressure.

Histograms are bar charts that show the occurrence of specific values from a channel. They are particularly useful for rpm, speed and throttle position, along with damper velocity. Comparing histograms from different sessions will give you an insight into the effects that changes to the car/driver are having. An rpm histogram tells us if the engine is being used within its power band for a high proportion of the lap. A speed histogram will tell us the characteristics of the circuit; by targeting the most-used speed ranges you will be able to make greater lap-time savings because a small increase in performance in this speed range

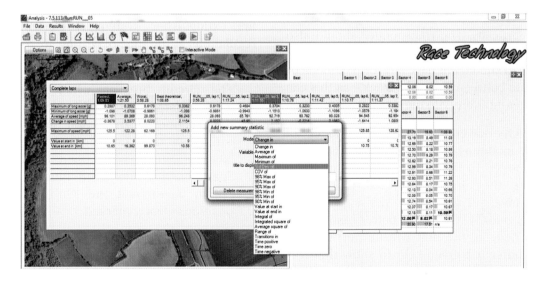

Figure 6.6 Channel reports

will make a bigger overall difference than to an infrequently used speed range. For example, a circuit with a set of high-speed ranges may benefit from an aero package with reduced drag and a stable chassis set-up, whereas a circuit with more low-speed ranges may have a greater need for traction, response and agility.

A throttle histogram will often look very different to the other histograms, as most of the lap time will be spent with the throttle at 100 per cent and the rest at 0 per cent. Most notably, the fastest lap achieved will often come from the lap when the throttle is at 100 per cent for the longest period (assuming a limited amount of wheel spin). This would show that the driver can stay on the throttle for longer before braking late for the corner and can apply full throttle very quickly on the exit of the corner. The area between full and closed throttle will tell us a lot about the car/driver/circuit. Improving the driver and car will result in a smaller amount of usage of the throttle between open and closed.

XY graphs provide a way of comparing one variable with another. For example, we can monitor oil surge by plotting oil pressure against lateral g-force. A friction circle can be shown by plotting longitudinal acceleration against lateral acceleration. A gear chart can be created from rpm against speed.

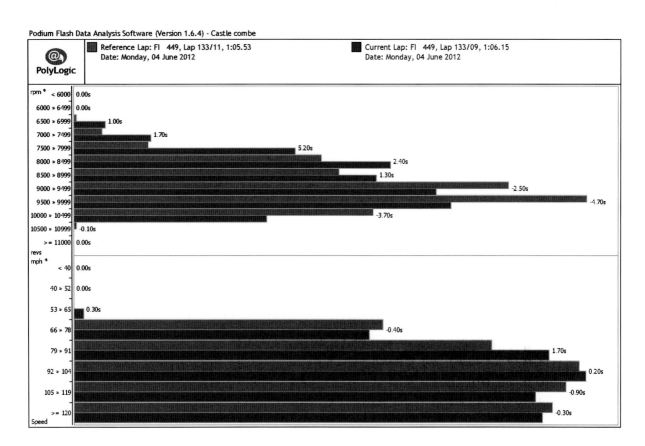

Figure 6.7 Histograms

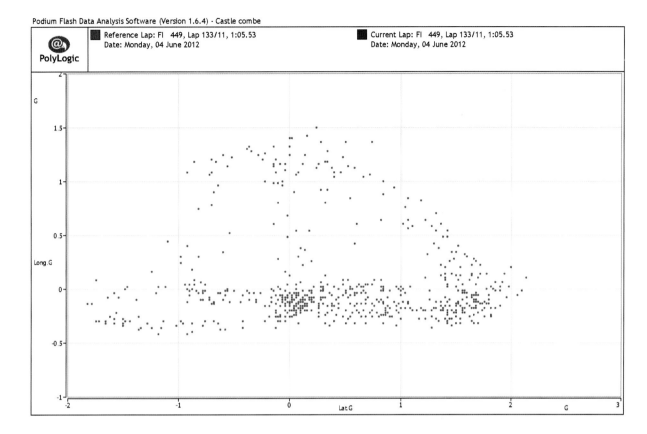
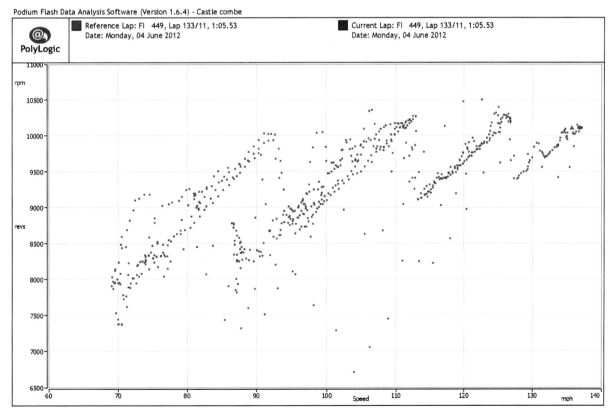

Figure 6.8 XY graphs

6.4.1 Driver data analysis

When analysing data, it is important to have a driver who can also provide good feedback as to how the car is working on the race circuit.

When viewing data on the software package, you need to break the data down into sectors; ideally every corner (including the braking, entry, apex and exit) should have its own sector. However, when analysing the data with the driver, these sectors must be broken down even further into stages, including:

- straight line initial braking stability
- entry to the corner, including gradually releasing brake pressure and steering
- corner apex, normally where the maximum steering angle can be observed, and the switch from brakes to throttle application
- corner exit, application of the throttle and reducing steering angle.

On an appropriate table and a track map (so as not to get confused about which corner is which) the engineer and driver can discuss how the car performs at every point of each corner to determine understeer or oversteer, braking stability and throttle application. This discussion is best had as soon as the session finishes and with video footage, if available, in order to relay fresh information – if an intercom is used, this is also a benefit.

From the collated comments, the software can be used to analyse the driving style and pinpoint certain areas to improve lap times. The cornering phase in an ideal scenario should be:

- maximum use of the tyre G-plot as discussed in Chapter 3 Chassis during the whole cornering phase
- maximum braking pressure at the beginning of the braking zone to make use of aerodynamic downforce
- a gradual release of pressure as the driver begins to turn into the corner
- maximum steering angle at the corner apex where the switchover point between brake and throttle occurs
- reduction of steering angle and throttle is then applied.

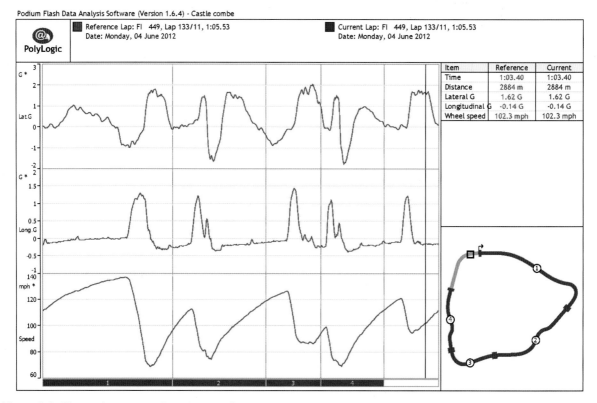

Figure 6.9 Channel traces to show corner phases

The quicker the driver can straighten the steering wheel and get hard on the throttle, the better.

It is also helpful to look at the characteristics of the driver's braking style, which can be viewed using a brake pressure sensor, or with the use of a longitudinal g-force sensor.

The speed trace can show the characteristic of the whole corner – ideally it should look like a V shape and not a U shape. Some long corners or those with a double apex may not produce a V shape but, on the whole, most should. This shows a steep deceleration via the brakes followed by an instant and hard switchover to the throttle.

The brake trace should show an instant peak of braking g-force, followed by a gradual bleed of the brake pressure. Any deviation from this could show that the driver is braking too early or not using the brakes well enough.

Driver lines can be analysed with corner radius and time slip. Corner radius can be found from:

$$\text{Corner radius} = \frac{\text{Speed}^2 \text{ (m/s)}}{\text{Lateral g-force (m/s}^2\text{)}}$$

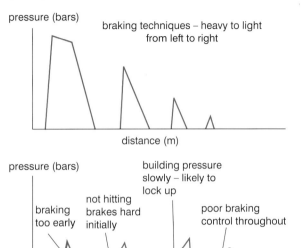

Figure 6.11 Braking traces

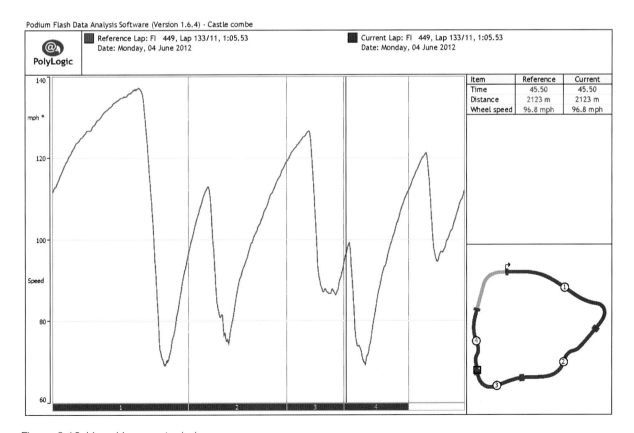

Figure 6.10 V vs. U corner technique

6.4.2 Chassis analysis

Comparison overlays are a simple way to analyse any set-up changes that have been carried out to the car, such as suspension and aerodynamic changes. Vehicle dynamics can also be analysed with the use of maths channels.

Longitudinal g-force can be used to analyse the performance of the braking ability of the car and acceleration of the car.

Lateral g-force can give a good indication of cornering performance analysis (even better when coupled with steering and throttle traces). The trace should show a high average g-force during the cornering phase, and a well set-up car should have a relatively square curve to show that the car is well balanced. A gradual build up to the peak lateral g-force usually indicates a reluctance of the car to turn into the corner, while any sharp and short dips in lateral g-force usually indicate oversteer that is quickly corrected by the driver with a correction of opposite steering angle. Understeer can usually be determined from a more gradual and longer dip in lateral g-force during the cornering phase.

Combined g-force is calculated using:

$$\sqrt{(\text{Lateral g-force}^2) + (\text{Longitudinal g-force}^2)}$$

Looking at a lateral and longitudinal g-force trace can allow you to see if the driver is reaching peak force at every corner and braking zone, while using the combined g-force (above) will show you where the driver may be able to be faster in a phase of the corner by any noticeable dips in the trace.

Steering angle usually uses a positive value for right-hand turns and negative values for left-hand turns. Steering angle can show understeer and oversteer, particularly when overlayed with lateral g-force.

Damper velocity histograms show how well the dampers are set up. In bump and rebound, the damper velocity should be symmetrical to show good control of the spring; any variation in the chart will show that an adjustment needs to be made to the damper settings, either bump or rebound (high or low speed, dependent upon adjustment range).

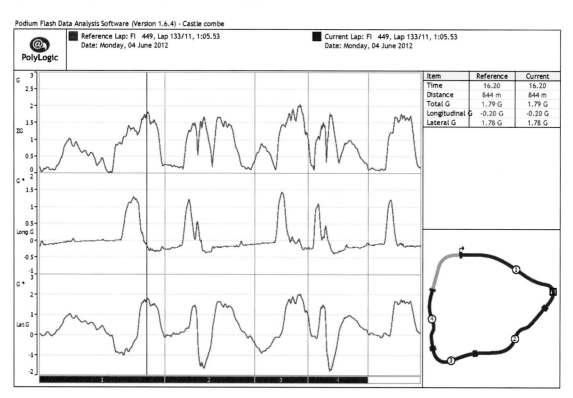

Figure 6.12 *Lateral, longitudinal and combined g-force traces*

Figure 6.13a shows an even amount of both bump and rebound velocities. This shows that the damper is generally set up well in both conditions. However, Figure 6.13b is asymmetrical with an uneven spread of damper velocities. This shows that the set-up is either poor, or there is another underlying issue with the set-up of that suspension corner. The damper should be dyno tested to further investigate.

Downforce can be assumed from using an XY graph to plot suspension travel against speed. As speed increases, we can expect the suspension to compress and, with a knowledge of the amount of compression and spring rate, we can assume downforce. If we have a spring rate that can be converted to N/mm, we can take that rate and multiply it by the suspension travel.

So if our equivalent spring rate is 10 N/mm and the spring has compressed 10 mm:

10 × 10 = 100 N of downforce

Total roll angle of the car can be calculated by using the following calculation:

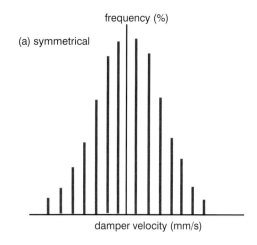

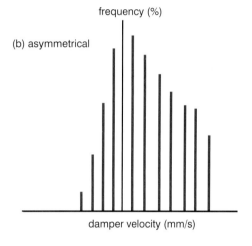

Figure 6.13 Damper velocity histograms

$$\text{Roll (degree)} = \arctan\left(\frac{((SLF - SRF) \times MRF) + ((SLR - SRR) \times MRR)}{TF + TR} \times \frac{180}{\pi}\right)$$

Where:

SLF = Suspension left front   SRF = Suspension right front

SLR = Suspension left rear   SRR = Suspension right rear

MRF = Motion ratio front   TF = Front track width (mm)

MRR = Motion ratio rear   TR = Rear track width (mm)

Suspension = Damper potentiometer travel (mm)

If we are only sensing damper travel, we simply create a channel that uses the motion ratio calculation that was discussed in Chapter 3 Chassis.

Plotting roll angle against lateral acceleration will show roll stiffness in degrees per m/s².

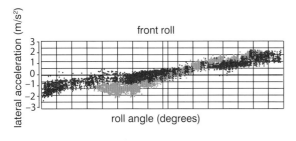

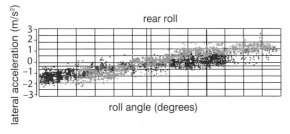

Figure 6.14 Roll stiffness

6.4.3 Engine analysis

Monitoring the water, oil and fuel performance of the engine is useful to ensure reliability, minimal engine wear and that the engine is fully up to temperature before being driven hard.

The rpm trace can provide information to help determine if the gearing is correct for the circuit or if the car has been over-revved. For example, whether the driver is up and down shifting at the right rpm, what the rev drop-off is after each gear change, and whether we are using the engine power band to the best of its ability. Plotting rpm against time, we can also analyse how well the driver changes gear, which is not so important for sequential or semi-automatic shifting, but is for H-gate type gearboxes.

The throttle trace will show how much throttle is being applied and how. We can see if the driver is applying the throttle promptly on the exit of the corner or see if they are applying it too early and subsequently having to lift off the throttle, wasting time. Coupling this trace with steering angle can show how the car behaves

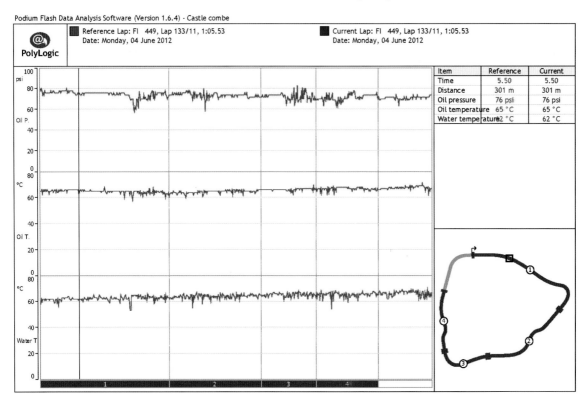

Figure 6.15 Engine oil and water performance

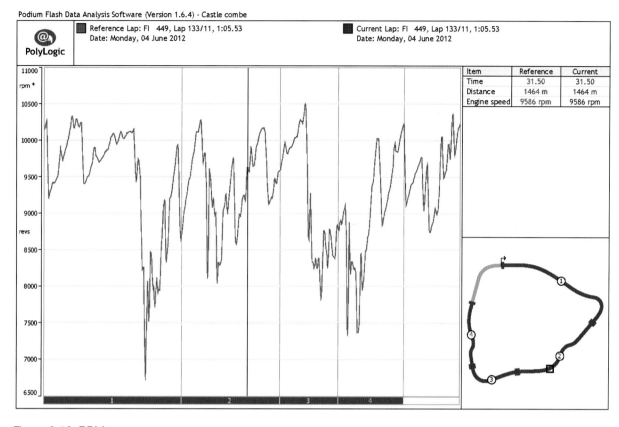

Figure 6.16 RPM trace

on corner exit when the throttle is applied. For example, if our steering trace shows a violent amount of oversteer occurring on corner exit, we can look at our throttle trace to see what the driver is doing. It may be that they are literally stamping on the throttle, which is upsetting the car's balance.

6.4.4 Maths channels

Maths channels are not physical channels connected to the data logger, but are manipulated channels created in the data logging software that give us the ability to perform mathematical equations on channel(s) in order to provide another useable outcome.

The software will carry the calculation out for us, we just need to tell it what to do with the raw data collated from the channels.

Simple calculations can be used to add, subtract, multiply and divide numbers and other channels; IF functions can also be used. You can also programme in your own constants so that you can look at particular areas of the vehicle's dynamics by pre-programming the wheelbase, track, gear ratios, tyre sizes, and so on. More advanced equations can also use differentiation and integration. Some examples for the uses of maths channels can be seen below.

For brake bias, we can use:

$$\text{Brake bias} = \frac{\text{FBP}}{\text{FBP} + \text{RBP}} \times 100$$

Where:

FBP = Front brake pressure

RBP = Rear brake pressure

Differentiation can be used to find the rate of change (distance, velocity and then acceleration).

Integration has the ability to show us the area under the curve and, therefore, how frequently an event occurs during a lap. The integral of a pedal position, such as the brake pedal, will tell us how long we are on the brakes during a lap.

IF statements can be used to determine wheel lock or brake trail.

When the differential is open there will be an evident difference between the rear wheel speeds. When the differential is locked the wheel speeds will be the same (within the accuracy of the sensors).

A maths channel can show us the operation of the differential by using the following calculation:

$$\text{Diff function} = \frac{\text{RLWS} - \text{RRWS}}{\text{Speed of car}} \times 100$$

Where:

RLWS = Rear left wheel speed

RRWS = Rear right wheel speed

CHAPTER 7
Basic engineering and preparation

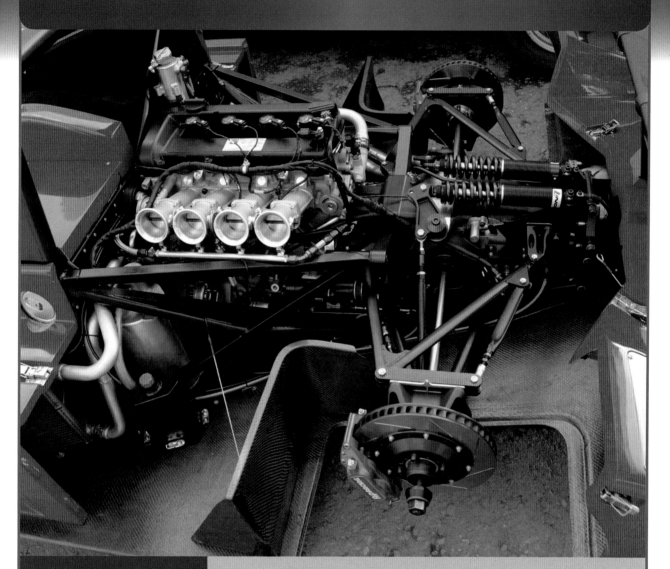

This chapter will outline:

- Preparing a car
- Checklists

7.1 Preparing a car

Nothing should be left to chance on a race car and, while outright speed wins races, reliability often wins the championship. When the car is being built, engineered, refreshed and prepared we must take a few basic things into account.

The two main aspects of the race car are the tyres (see Chapter 3 Chassis) and Newton's second law of motion:

> Force = Mass × Acceleration
>
> Or
>
> Acceleration (m/s²) = $\dfrac{\text{Force (N)}}{\text{Mass (kg)}}$

Rearranging the equation as above shows that the higher the force and the lower the mass, the more acceleration we can potentially produce longitudinally, laterally and when combined. (**Note:** $1\,g = 9.81\,\text{m/s}^2$). Although the mass does not change, the force for each of our accelerations will change to engine power, braking power and downforce. The tyres, which are discussed in detail in Chapter 3 Chassis, are the key to unlocking the potential acceleration available so, quite simply, the performance of the tyre will determine our lap time above everything else.

We must also remember that any rotational or reciprocating inertial mass needs to be accelerated, and so the lighter these components are the better (so long as reliability is not compromised); this includes the engine, transmission, wheels, tyres and brakes.

Compliance and stiction in the steering and suspension is an area that needs to be thought about as this can make setting the car up difficult. It can reduce the feedback that the driver needs and will make the car insensitive to set-up changes. Rod ends and spherical joints not only reduce compliance and stiction, but also provide a finer amount of adjustment. Although polyurethane bushes are used in saloon cars and similar, you will rarely see them on a pure race car. The joints come in different qualities and, therefore, prices. If you convert a car from bushes to joints, however, where will all the vibration and loadings go and what will absorb them? In this situation, additional stress is placed upon the rest of the suspension and chassis components, which will lead to increased levels of fatigue, increasing the chances of failure.

A race car needs to be well built and prepared in order to successfully compete at events and to cause as little stress to the team as possible during the event. The car and all its systems must be prepared and maintained to prevent any failure before it occurs. To finish first, first you must finish!

Preparing a car often takes time but the basics of it are usually simple. Keeping a clean car is the very start of race car preparation – the more thorough you are with this process, the more likely you are to find any issues with the car. A chassis with dirt and debris in the corners and in hard-to-reach places can easily disguise cracks in the chassis and surface rust, which causes pitting and weakens the material. Wiping all the parts of the car with penetrating oil will help to clean the car and protect it from moisture and contaminants.

7.1.1 Fasteners

Once clean, a full nut and bolt check should be carried out to ensure nothing is working its way loose. Torque Seal is an excellent product that can be placed on the outer edge of the bolt thread and nut face and, when applied, it sets. As soon as anything comes loose, the Torque Seal will fall off, so you can easily identify anything that needs attention. It comes in a variety of bright colours so that it can be visible in any application. It may sometimes break off when exposed to high amounts of vibration; however, it significantly cuts the time taken to do a full nut and bolt check, and reduces the chance of someone continually over-tightening every nut and bolt on the car.

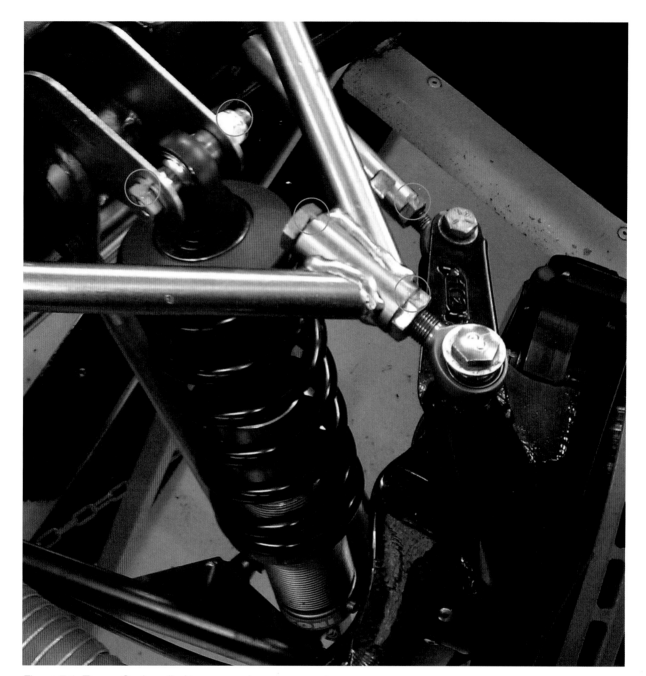

Figure 7.1 Torque Seal applied to suspension components

Methods of fastener security can include thread lock, lock wire and split/roll pins, along with the normal range of locking nuts and washers.

Thread lock comes in a variety of specifications that will determine temperature range, breakaway torque, material type, finish and size. It is applied to the thread before being fitted and, once dried, will stop the fastener from coming loose. Lock wire is another form of additional fastener security that can be used for sump bungs, engine seals, carburettor parts and suspension parts, such as bolts for upright pivot points. When using lock wire, ensure it is applied in the right manner – you must wrap the lock wire in the direction of tightening of the fastener (i.e. clockwise for conventional right-handed threads).

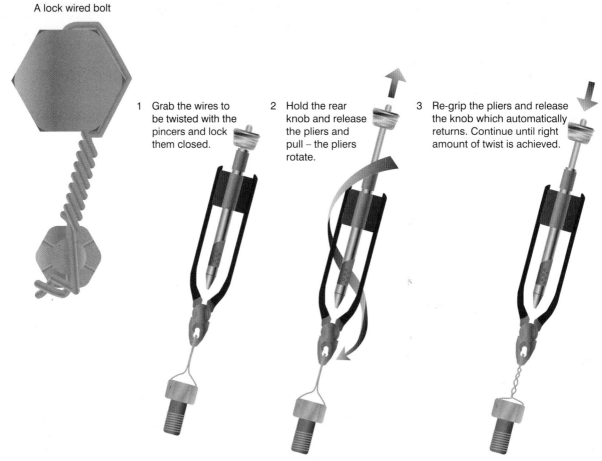

Figure 7.2 Lock wire and lock wire pliers

Types of nuts and bolts

Nuts and bolts are most commonly imperial (UNF – unified coarse; UNC – unified fine) or metric. Imperial is very common on race cars, in particular, with suspension including rod ends. Metric bolts are identified with the following code example: M10×1.5×40. This indicates that the bolt diameter (not the head size) is 10 mm, the thread pitch (distance between each thread) is 1.5 mm and the length of the bolt (not including the head) is 40 mm. Metric bolt strength is determined by its tensile (stress level at which the bolt will break under a tensile (pulling) force) and yield strength (point at which the bolt permanently deforms after reaching the elastic limit). You may have also heard of torque-to-yield bolts, which are used for cylinder head and big end bolts; these are actually tightened beyond their yield point and stretched. On the face of the metric bolt head will be a number that identifies its strength properties. A steel bolt will usually be 8.8, 10.9 or 12.9. Steel bolts can be cadmium or zinc plated to protect them from corrosion.

To show the difference between two bolts, the strengths can be seen below:

- 8.8 – a tensile strength of 800 N/mm^2 and a yield strength of 640 N/mm^2
- 10.9 – a tensile strength of 1040 N/mm^2 and a yield strength of 940 N/mm^2

Imperial markings on the bolt head tend to be a number of lines radiating from the centre of the head. You will find that a lot of race cars use imperial (UNF) fasteners for the suspension. This is mainly because rod ends generally come from the USA and have a wider variety of size availability when compared to metric.

When using imperial joints on the suspension and steering, it makes sense to fit the whole suspension together with imperial fasteners. AN (Air Force–Navy Aeronautical Standard) and NAS (National Aerospace Standard) series fasteners are of the highest quality, are made with a higher-quality material, and come in a wider range of options, in terms of shank and thread lengths, which is handy for ensuring that we have a full shank length on our suspension bushes for instance. Using NAS or AN bolts with K-nuts ensures an optimum fit, high strength and low weight (see Figure 7.4 for bolt terminology).

A bolt with three radiating lines from the centre to the outer edges is a Grade 5 bolt made from a medium carbon steel, which is quenched and tempered with a tensile stress rating of a minimum of 120 ksi.

A bolt with six radiating lines from the centre to each edge is a Grade 8 bolt made from a medium carbon alloy steel, which is quenched and tempered with a tensile stress rating of a minimum of 150 ksi.

1000 psi = 1 ksi
1 ksi = 6.895 N/mm²

Table 7.1 shows some common imperial sizes, and their conversions into metric (to 3 decimal points).

Stainless steel fasteners have the benefit of being corrosion resistant, giving low maintenance. They also have a higher surface-to-surface coefficient of friction, which means the clamping force will be less than a steel bolt at the same tightening torque. They also weigh more.

More exotic materials can be used for a variety of purposes. Companies such as Pro-Bolt offer fasteners in aluminium and titanium, which offer benefits such as being lightweight and, in the case of titanium, still very strong.

Figure 7.3 Lightweight drilled titanium bolt

Table 7.1 Imperial, decimal and metric sizes

Imperial size (inches)	Decimal equivalent	Metric equivalent (mm)
1/4	0.250	6.350
5/16	0.313	7.938
3/8	0.375	9.525
7/16	0.438	11.113
1/2	0.500	12.700
9/16	0.563	14.288
5/8	0.625	15.875
11/16	0.688	17.463
3/4	0.750	19.050
13/16	0.813	20.638
7/8	0.875	22.225
15/16	0.938	23.813
1	1.000	25.400

Set screws or bolts?

A set screw has a thread from the underside of the head, all the way to the bottom, while a bolt has a plain shank under the head, and then a thread after the shank. Fixing load-bearing components together should be carried out with a bolt, where the shank supports the two components and stops any lateral movement from occurring. You can also get different designs of head, including hex, socket and pan head.

Once we have decided on the bolt type, it needs a nut to secure it and, most likely, a washer to spread the load. For example, we can use nylocs, K-nuts, castellated nuts or lock nuts, all for different applications.

Nylocs are the most commonly used, where the nylon insert grabs the thread and the high coefficient of friction stops it from coming loose. However, the nylon becomes damaged when it is used over and over again, so will need to be replaced eventually. In a high-temperature environment, the nylon can melt and cause the fastener to work loose, something I have heard of (but thankfully never experienced) on three-piece wheels, whereby the heat of the brakes actually melted the nylon and the wheels started to fall apart.

Figure 7.5 Nyloc nuts (**Note**: a blue insert is metric and a white/clear insert is imperial)

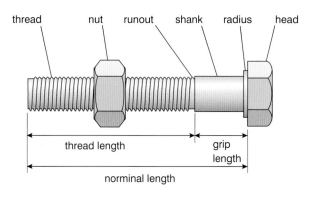

Figure 7.4 Set screw and bolt

The K-nut is another fastener from the aerospace industry that can be found in both metric and imperial form. They work by having a deformed elliptical shape instead of a round hole, and so the bolt has to force the nut into a round shape in order to tighten up or undo itself. K-nuts can handle high temperatures and can also be reused. The hexagon shape around the outside and, therefore, its thickness, is a lot smaller than a conventional nut so they are much lighter – and more expensive.

Figure 7.6 K-nuts

Washers can also be used, not only to stop fasteners working loose, but also to spread the load on the clamped components. Spring washers, tab washers and star washers can all be used as locking devices.

As a general rule of thumb, a formula of 1.5 times the bolt diameter is the minimum amount of thread that should be inserted into female thread (i.e. an M10 bolt should have 15mm of thread inserted into its female counterpart).

When fitting a nut and bolt, if there is an option to position vertically, it would be advisable to place the bolt in through the top with the nut on the bottom to act as a failsafe, so if the nut did come loose and fall off, there would be a chance that the bolt would remain in place.

Other fasteners that are commonly used:

- Anchor/captive nuts – these can be either fixed or floating in their housings. These are great to use in areas where it is impossible or difficult to get to one side of the nut or bolt. The threaded bush section is fixed to a small plate that will be fixed to one of the parts that needs fastening together. You can buy these in the form of plain nuts or with a locking K-nut type bush.
- Cable ties/anchors – these are ever popular in motorsport and often used for tidying up wiring looms and fluid lines. Anchors can be riveted to chassis and body parts to allow the cable ties to make things even neater. They are most commonly made of PVC and are available in various sizes and colours. Stainless steel tie wraps are also good for securing CV boots and securing heat wrap to exhausts.
- Dowels – these are used to allow you to precisely line up two components with each other every time that they are separated.
- Dzus fasteners (1/4 turn) and cam locks – these quick-release fasteners are useful for panels, such as bodywork and inspection hatches. The fastener will have a slot in its front face to allow it to be turned and a keying device at the other end that locks into a simple mechanism on the inner panel.
- Heli-Coil/thread inserts – these clever tool kits allow you to create a new thread if you have stripped or cross-threaded something. All you have to do is get a Heli-Coil of the same thread type that you have stripped, drill the damaged thread out to the correct diameter given to insert the Heli-Coil, use the correct tap to produce a thread for the Heli-Coil to sit in, wind the Heli-Coil into the part with the tool and then knock the tab off the end of the insert once wound in. You then have a new thread of the same size and pitch as the previously damaged thread.

- Lock nuts – these half-width plain nuts are the only solution that should be used for locking a rod end into place. Once set to the correct length, the lock nut should tighten the thread of the rod end up against the end of the female thread. It is normally best to avoid thread lock and lock wire for this scenario as you will frequently be altering the set-up of the car, so some torque seal and a regular visual inspection is the best option.
- Nappy pins and R-clips – these are used mainly for secondary retaining devices of centre lock wheels nuts, which prevent the wheel nut completely falling off if it ever became loose.
- Over centre clips – these are used for joining large pieces of bodywork together and can provide a good clamping force to safely secure two panels together. I prefer to use the clips with some form of adjustment screw in them to allow the clamping force to be set by the user.
- Pip-pins – these pins are essentially a tube in a tube with two detent balls at one end. When the pin inner is pushed or pulled, the balls retract and allow panels to be securely fastened. In its normal state the inner pin keeps pressure on the back face of the balls to stop them from being removed.
- Riv-nuts – these threaded inserts allow you to place a thread into sheet metal or tube/box section walls and even composite panels. They are fitted with a tool that threads into the insert and allows you to squeeze the outer part together so that it clamps against the sheet. Over time, they may vibrate loose as they lose their grip on the panel. They can be used for a variety of purposes, such as inspection hatches and holding lightweight components to various parts of the car (e.g. fuel pumps and other electrical equipment).
- Rivets – these permanent fixings come in a variety of shapes, sizes and materials. They are commonly used in race cars, particularly for fixing panels to the chassis, for aluminium wings along with holding various brackets to the chassis. Vibration, along with any chassis flex, can often loosen a rivet. When lining up panels to rivet to a chassis, you can buy rivet clamps (Cleco fasteners), which are semi-permanent

Figure 7.7 A range of fasteners

rivets that allow you to temporarily fix the sheet in place, making the alignment of the panels much easier.
- Rubber clamps, blocks and nylon – these are often used to help absorb vibration and protect electrical circuit boards, and so on. Nylon is also easily machinable and can be used as a spacer or bush on non-safety-critical parts to save weight over using a metal equivalent.

7.1.2 Component management

Brackets, plates, fixings and so on should always have radiused edges, to prevent any concentrated stress areas that could fracture, and should have no sharp edges. If any cracks appear and the part cannot be replaced at the event, you should drill a hole at the end of the crack to stop it creeping any further and then address the crack appropriately, depending on the material used.

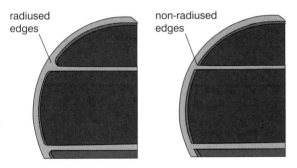

Figure 7.8 Radiused and non-radiused edges

Life charts

Life charts or logs allow you to keep track of how many hours or miles each part of the car has been used for, and will allow you to refresh or replace parts before they become worn out, damaged or fail. There is no limit to which parts can and cannot be 'lifed'. (By lifed, we mean specifying a duration of time that a component can be used on a car before it reaches a period where it should be inspected, serviced or replaced.) Of course, there will be parts of the car that do not necessarily require life charting, but anything under load or stress will need to be paid the most attention (brakes, suspension parts, engine, transmission, aero parts, etc.). Unless they are critical safety parts, most parts that have reached their life limit can be kept as emergency spares or for testing purposes.

Heat management

You may find that areas of high heat, in particular the exhaust system, will need to be shielded from vulnerable parts, or those parts will need to do the shielding. On the exhaust side, this could include heat wrap, high-temperature coatings or shields; while protecting the parts themselves can include heat wrap tubing, gold film or foil and ensuring pipes and wiring are routed away from heat sources if possible. Temperature stickers can also be placed on various components, such as brake callipers, underfloors or cylinder heads, to monitor maximum temperatures.

exhaust heat wrap

temperature sticker

gold foil

heat sleeving

Figure 7.9 Heat management methods

Protective surface finishes

With all race car components, parts, fasteners and so on, we should be looking to maximise their life expectancy by ensuring that they are adequately protected from their surrounding environment. While heat management, lifing charts and penetrating oil all make a difference, protecting the actual surface finish of the parts means we are not constantly addressing damage to surface finishes and corrosion after every event. In wet weather racing, water, dirt and grit will find their way in to every part of the car, so it is essential to remove the water and clean the car thoroughly – choosing a long-lasting surface finish will help us protect the longevity and appearance of the car.

The main options include painting, plating and anodising:

- Painting – this is obviously used for the bodywork of the car, although with most composite-covered cars, a coloured gel coat is applied before the structural lay-up of the composite layers begins. The suspension parts, chassis and associated panels and most brackets will often be powder-coated, and will give a long-lasting and durable finish.
- Plating – this usually comes in zinc, cadmium, nickel and chromium finishes. These platings provide corrosion resistance to steel components. The thin nature of the plating means that there are no issues with threads, bushes and sleeves needing to be filed out after the coating process (which powder coating often requires). These platings also make it easier to spot any cracks in components with a visual inspection than with powder coating. Figure 7.10 shows zinc-plated and yellow passivate suspension links.
- Anodising – this is used to protect aluminium components. This surface coating is an electrolytic process similar to plating, which provides no change in dimensions to the component being coated. It is available in a range of different colours and has excellent resistance to corrosion.

Figure 7.10 Zinc plating

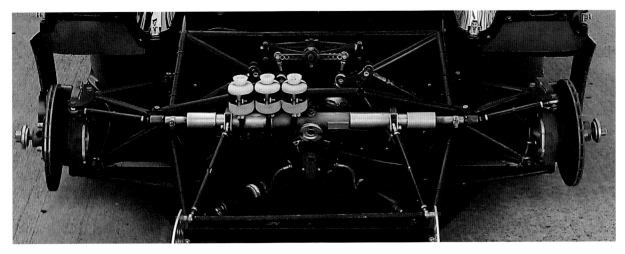

Figure 7.11 Anodising (see purple suspension links)

7.2 Checklists

Producing a laminated checklist (job list) for preparing the car means it can be wiped clean after the preparation has been carried out. It is best to put an initial and date box next to each item to check so you can see who has carried out the task and when it was carried out.

Checklists will vary depending on the car in use, type of event, and where and when the checklist will be used (pre-, during or post-event).

The pre-event checklist will include the usual jobs to maintain and prepare the car, as well as setting the car up for the next event based on previous data gathered, predicted weather conditions, any driver/car preparation specific to the event and any replacement parts that may have reached their life limit.

On event, you would again carry out maintenance tasks, but they would be limited as you may not transport the same equipment to the event as what is in your workshop. The additional duties for on event include tyre and fuel management, as well as completing run sheets and set-up sheets for each session so that you can monitor set-up changes and performance over the course of the event.

Post-event, you would look at creating reports and analysing the weekend in terms of car and driver performance and the efficiency of team operations. Lifing and mileage should be totalled and recorded. The team and drivers should be debriefed on the event and any concerns discussed, followed by analysis of the reasons why the concerns came about and what can be done to eliminate them. You will normally expect an additional job list from the event, which may determine stock levels in the race truck for consumables, additional tools required or additional areas of the car that need attention once it is back in the workshop.

Figure 7.12 is one of the checklists that we use while on event when racing at the college.

As already discussed, the checklist will depend on the car type as they are all built in different ways with different systems, while the type of event often dictates how much time you have between each session. The duration and type of competition event will also determine what will be checked between each session.

As you can see from the checklist example, a wide range of checks must be carried out between each session, but ensuring all of the boxes are ticked will enable the car to stay reliable, competitive and presentable. All of the checks should be self-explanatory and straightforward.

196 Basic engineering and preparation

Race preparation checklist		
Event: Venue:	**Date:**	
Task	**Checked**	**Technician**
Inspect and clean chassis, body and suspension components	☐	___
Check brake and clutch fluid levels	☐	___
Check oil level	☐	___
Check coolant level	☐	___
Check wheel nuts	☐	___
Check gearing correct for circuit	☐	___
Check chain tension and alignment	☐	___
Lubricate chain	☐	___
Full nut and bolt check (visual inspection of torque seal) incl. wheel bearings	☐	___
Check engine for any fluid leaks	☐	___
Check radiator and oil cooler for damage or blockage (cover to suit temp.)	☐	___
Check throttle cable for damage or free play	☐	___
Check air filter	☐	___
Check dash and data logger operation	☐	___
Drain fuel and refill with required amount	☐	___
Slacken harnesses	☐	___
Check fire extinguisher pin (remove before car goes on track)	☐	___
Check wheel weights and cover in tape	☐	___
Adjust tyre pressures and check tyre condition	☐	___
Check bodywork clips and cover in tape	☐	___
Check mirrors are secure, adjust with driver in car	☐	___
Check splitters and wing are secure	☐	___
Suspension adjustments completed	☐	___
Aerodynamic adjustments completed	☐	___
Camera check, security, card, clean lens	☐	___

Figure 7.12 Sample race preparation checklist

7.2.1 Simple preparation tasks at the workshop

- Make sure all batteries are charged, including race car, jump batteries, beacon and truck batteries. Always use the correct charger for your battery.
- If possible, cars should be put in chassis stands so all checks can be carried out.
- Before going to the next event, make sure the tyres have more air in than required as it is easy to take air out at the track and makes the car easier to push around.
- Check fluids and filters.
- Always torque up wheel nuts. If your car has a single wheel nut, NEVER put the wheel pins in until the wheel nut has been done up. This means that if there is no wheel pin, you know that the nut has not been done up.
- Make sure you have enough fuel for the event.
- Check that all the packing is in the silencer so you do not fail noise testing.

7.2.2 Simple track tasks

- Set tyre pressures.
- Start the car and warm the engine to within 10 °C of running temperature.
- Check your tick list.
- Get the car ready at least 10 minutes before it has to be on track.
- Tape up radiator and oil cooler ducts if required.
- Fit test day or scrutineering sticker.
- Set up the beacon if you are using a logging system.
- Keep car clean at all times.
- Get a cup of coffee and a breakfast roll (or whatever you like). This sounds strange, but it is easy to forget to eat and drink on a busy weekend and when you become hungry and dehydrated you could lose concentration and make mistakes.

7.2.3 Things that you must do

- Always take care in the pit lane as it is a dangerous place.
- If possible, wear gloves when working on the car.
- Always take a stopwatch.
- Remember, race tracks are cold places so always take warm clothing and shoes or boots.

Component 'lifing' chart

The chart gives the recommended life expectancy of components under 'normal, on-track racing conditions'. If some of your racing time is done 'off-track' or you hit kerbs, pot holes or other cars – hard, then you will need to consider reducing the timescales recommended. On the other hand, more 'gentle' trackday use will obviously extend the recommended time!

Component life chart

Component	Hours	Action
Engine	30	Rebuild
Gearbox	30	Inspect/rebuild
Chain	10	Replace
Suspension bushes	50	Replace
Suspension rose joint	30	Replace
Front upright including hub	90	Replace
Front wishbones	90	Inspect/replace
Rear upright	90	Replace
Rear hub	90	Replace
Rear wishbone	90	Inspect/replace
Drive shafts	60	Replace
Calipers	60	Rebuild
Brake discs	10	Inspect/replace
Shock absorbers	60	Rebuild
Steering rack	60	Inspect/rebuild or replace
Brake master cylinder	60	Inspect/replace
Wing support stays	60	Inspect/replace
Fuel tank		Remove and inspect annually

Figure 7.13 An example of a component life expectancy chart

CHAPTER **8**

Event and set-up

This chapter will outline:

- What happens at an event
- Getting involved in motorsport
- Set-up

8.1 What happens at an event

A race event can be a stressful and hectic experience, but this can be avoided with the correct preparation and planning prior to the event, and good communication and teamwork during the event.

All race events run in a similar fashion, although they vary slightly from circuit to circuit and from race organiser to race organiser. A check of the final instructions that will be available prior to the event should confirm any details that might be different to other race events. Always ensure you also have a current copy of the MSA Blue Book (officially known as the *Motor Sports Association Competitors' and Officials' Yearbook*) to ensure you can easily refer to any rule or regulation should you need to.

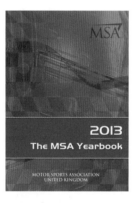

Figure 8.1 MSA Blue Book

Prior to arriving, you should ensure that you have the final instructions for the event, entrance tickets, the event timetable and a paddock plan. On arrival you should park in your allocated parking area (unless you are testing, when you may have booked a garage). Everything should then be set up ready for the weekend, including the awning, flooring, wheels and tyres, refuelling equipment, tools, spares and, of course, the car.

8.1.1 Items to take

The tools and spares that you carry completely depend on the car and event type, your budget, and how much you can safely carry in your method of transport.

Important spares include the following:

- Wheels and tyres
- Fuel
- Electrical repair kit (wires, connectors, fuses, batteries, etc.)
- Spare fasteners, including wheel nuts, bodywork fasteners, rivets and cable ties
- Fluids – oil, water, brake fluid (including the method of refilling)
- Filters – oil and fuel
- Drive shafts, wishbones, rod ends, etc.
- Gears
- Water/fuel hose and clips
- Tapes and torque seal
- Exhaust wadding
- Throttle/gear/clutch cable
- Battery
- Brake pads
- Engine spares (spark plugs, coil, electronic control unit and sensors)
- Bodywork/wings/nose box
- Silencer

Important tools include the following:

- Jacks and stands
- All paperwork
- Refuelling equipment
- Brake bleeding bottle
- General tools, including spanners, screwdrivers, pliers (including circlip pliers), sockets, torque wrench, hex keys, mole grips
- Impact gun and sockets, drill and drill bits, rivet gun
- Hacksaw, emery cloth, files, knife
- Steel rule, tape measure
- Torch
- Chassis alignment equipment (corner weights if possible)
- Lock wire and tool
- Compressor, air line, tyre inflator, foot pump
- Battery jump pack
- Generator
- Pit board
- Paper towel
- Cleaning equipment for the car
- Sprays, adhesives and lubricants
- Drain pan
- Fire extinguisher

- Gloves and other personal protective equipment for working on the car
- First aid kit
- Pit board and numbers
- Tyre pressure gauge and tyre temperature gauge
- Whiteboard, clock, pen and paper
- Tyre marking pen
- Battery charger
- Multimeter
- Fuel drain-out kit
- Damper adjustment tools
- Brake cleaner and spray bottle
- Sweeping brush
- Bin
- Magnetic tray
- Stopwatch and relevant run sheets
- Umbrella
- Laptop computer
- Car camera
- Lap beacon and battery
- Pit trolley
- TV to use on the pit wall to monitor lap times
- Credit card for emergency purchases over the weekend

The layout of your set-up should allow easy entry and exit for the car(s) with adequate shelter and space to house everything needed for the event. Although the car should be fully prepared on arrival to the circuit, you should carry out some simple final checks on the car, such as a basic visual inspection, fluid levels and wheel nuts – particularly if you swap your travel wheels for the testing set.

Figure 8.2 Team work area at an event

8.1.2 Testing and noise

Testing is usually carried out on a Thursday or Friday and is a great opportunity for some track time. Getting the most out of the test sessions requires some planning. Driving around the track is fine for some practice laps and track familiarisation, but you should set some targets for testing. You will need to sign on and noise test separately for the test day and race event. Noise test levels can vary based on circuit and event type. At the time of writing, the regulations on noise for the MSA are:

> Measurements will be made at 0.5 m from the end of the exhaust pipe with the microphone at an angle of 45 degrees from the exhaust outlet and at a height of 0.5 m to 1.0 m above the ground.
>
> Section B Race cars (single-seater and sports racing cars) must be tested static at 3/4 rpm and must not exceed 108 dB (decibel).
>
> *2012 Competitors' and Officials' Yearbook,*
> Motor Sports Association, UK

Noise is a major issue in UK motorsport, and circuits, such as Castle Combe and Thruxton, are closely monitored and restricted. The balancing of noise issues with the wishes of the local residents is a source of difficulty for the motorsport community.

8.1.3 Planning and test sessions

The test day will either be split into sessions or will be in an open pit lane format; a timetable for the test day will be given out at the sign-on office. I find it useful to have a whiteboard or similar with a clock attached to it where I can write session times and other important information. This should be used, along with a checklist for the car, during the event. Intercoms can only be used in events with pit stops or when they take place over a long distance, but can still be used for testing. However, they can also be useful for communication across the circuit, pit lane and paddock area.

The first test session can generally be used for some basic citing laps to familiarise the driver with the circuit and the conditions, braking points and lines. As we know, the fastest way around a corner is to follow the route that straightens out the corner as much as possible (the less steering input you put in, the more throttle you will be able to apply). As a result, you need to use as much of the track as possible and this often includes the kerbs around the entry, apex and exit of the circuit.

I often find it best to walk the circuit before an event, as this allows me to easily notice gradient and camber change of the circuit, changes in surface type, and also how aggressive the kerbs are and whether or not my car will be able to attack them. For example, most kerbs at Silverstone are very flat and easy to attack, while the final chicane at Thruxton must be avoided at all costs in a sports racer or single-seater as it is very tall and could cause major damage to a car with a low ride height.

After the first session, feedback and data should be considered from the driver and the data logger. The remaining sessions for the day should ideally look at specific targets, such as lap time, fuel usage, reliability, bedding in new parts, testing new parts, trying new car set-ups and continual track/car familiarisation.

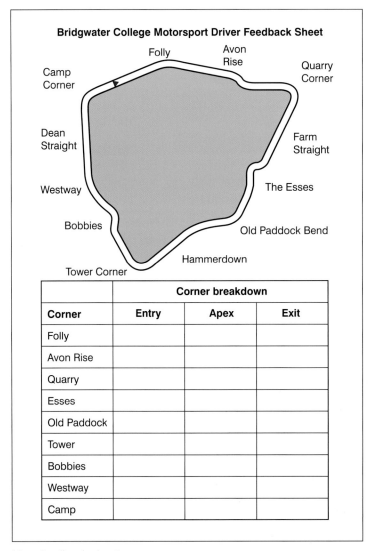

Figure 8.3 Example of a driver feedback sheet

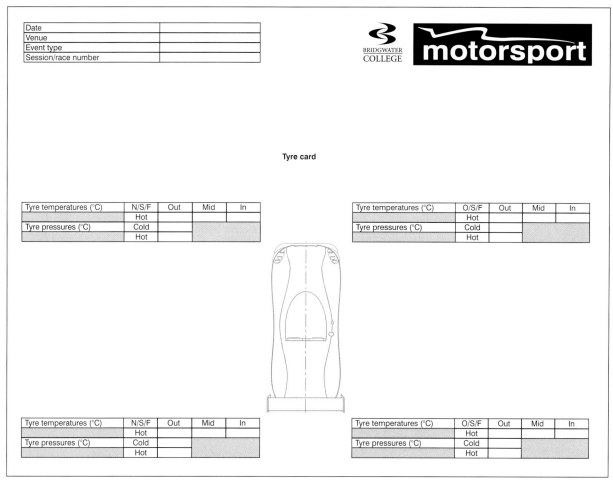

Figure 8.4 Example of a tyre information sheet

In terms of fuel usage, fuel weighs approximately 0.73 kg per litre. Carrying extra fuel is costly and will hamper braking, acceleration and cornering. Race teams and car manufacturers spend lots of money to reduce the weight of a car; finishing the race with a minimum of 3l is a mandatory requirement for fuel sampling and a dry break coupling must be fitted to accommodate this. If you finish the race with 10l more fuel than you need, this is an extra 7.3 kg.

Run sheets that contain lap times, set-up changes, session number, weather and track conditions, and driver comments should be kept for every session so that a log can be created for future use and analysis. This may show different aerodynamics configurations, damper sweeps, and so on.

In the last session (dependent upon budget), you may decide to lightly scrub in your new tyres for the event ahead.

In between each session, the event maintenance checks need to be carried out; this includes fluid levels, tyre conditions, visual inspections and so on. See Chapter 7 Basic engineering and preparation for a sample checklist.

At the end of the test day, a more comprehensive and thorough set of checks should be carried out to ready the car for race day.

8.1.4 Race day

Qualifying and scrutineering

On race day, the car will again need to be noise tested, the driver will need to sign on (and receive the scrutineering pass) and the car will have to go through a scrutineering check. This check will include general safety features of the car and its mechanical state. Always check that the rear lights work before scrutineering. The driver's safety equipment will also be checked. Specific series-related regulations may also be used to check the car's compliance to these. If the driver is new to the circuit, they will also have to attend a first-time drivers' briefing.

Qualifying will shortly follow scrutineering, unless you get a free practice session. In qualifying, you must complete three laps of the circuit in order to be able to be on the grid for the race. Depending on the format of the weekend, getting a good lap time during the session is obviously important, but a good second lap also often counts towards your starting position in race two (unless your finishing position from race one is taken as your starting position for race two).

Track conditions may have changed overnight so it is also important to provide feedback to engineers when returning to the pit lane or over the intercom, so that changes can be made for the race. A stopwatch in the dash display system is useful to indicate to the driver how much time is left in the session, along with a lap delta feature to show progress over the course of a lap.

It is important not to get involved in a battle with other drivers or sit behind other cars for too long (unless using them as a tow). You need to find some clear air in front of you and produce the fastest lap possible.

After qualifying, the times will be released showing the fastest lap for each driver and the grid for the race. Timing screens can be used during each session on the pit wall to allow teams to convey information to and from the driver about their progress during the session. A pit board should be used to relay this information. It is a good idea to have something on the pit board that allows your driver to easily identify their board. The pit board can show current position, lap time, lap/time remaining and the gap between the car ahead or behind. It is normally the driver's decision as to what is displayed on the pit board.

The race

The race itself will follow a similar process as qualifying and you will be called to the assembly area around 30 minutes before the race start time. You will then be assembled on the grid and have a green flag (warm-up) lap to produce some heat in the tyres, brakes and engine. This is not always the case with cars fitted with treaded road tyres. Once cars are back on the grid, the starter will display the countdown boards before the five red lights are illuminated all at once or one after the other. Once they are extinguished, the race begins. At the end of the race, you will have a slowing-down lap and will be guided back into the pit lane and directed either to the paddock/garages or to parc fermé where post-race checks can be carried out.

Note: no team members are allowed into parc fermé.

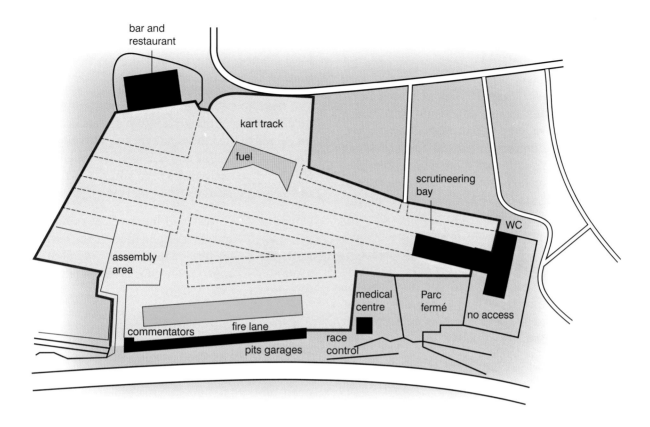

Figure 8.5 A paddock plan

8.1.5 Other considerations

Teamwork

This is essential no matter how many people are in the team. Most clubman race teams rely on friends and family for help – quite simply, the more hands available will make the work easier. Many teams rely on motorsport students to help them and then have a core experienced set of engineers and mechanics.

Timekeeping and good housekeeping

Both of these are vital for the weekend to run smoothly. Good housekeeping should be applied both in the workshop and at the event. Storage boxes, tool trays and magnetic trays are ways of ensuring that everything stays in an organised manner and does not get lost, misplaced or put back in the wrong place. Labelling items, sketching diagrams or taking a picture will take seconds and save vital minutes when reassembling systems on the car.

Spares and tools, storage and note taking

You can never have enough spares or tools and will be continually adding to the kit that you already take with you. It is a good idea to note down what stocks are getting low, what you needed on the event but did not have, and what could be better organised in the lorry/trailer so that it is easy to unpack and find.

Organisation

The layout of your pit area, the operation of pit stops, preparation and analysis need to be well organised, with each team member knowing their role, what to do and who to speak to if they have any issues.

8.2 Getting involved in motorsport

There are many different types of motorsport, including circuit racing, hill climb and sprint events, rally, rallycross, autotest, drag racing, karting, trials, motorbike racing and stock car racing. All have different rules and regulations but will be mainly governed by the Motor Sports Association (MSA). The MSA is the governing body for motorsport in the UK and is the only body that can issue race licences to competitors. If competing is not possible for you, then volunteering to help a team or individual is a good way to get hands-on or marshalling experience. This is also an excellent way of finding a job in motorsport as teams will realise your value very quickly.

Getting involved in competing at circuit racing requires some forward planning and thought as to what you want to do. The basic route into it is as follows.

Firstly, you need to obtain a 'Go Racing' starter pack from the MSA. At the time of writing, the Go Racing pack costs £59 and includes an introductory DVD, the MSA Blue Book, all of the necessary paperwork, and information about where you can take your Association of Racing Drivers Schools (ARDS) test.

8.2.1 The ARDS test

To gain a National B (standard for UK club racing) licence, you need to pass a practical and written test, known as an ARDS test. This test can be carried out at most major UK race circuits around the country. It lasts for approximately half a day. The practical test is not looking for outright speed but for awareness on the circuit, smooth driving, and the ability to pass other cars and be passed safely, and, of course, to show that you can carry an appropriate speed for the test. The written test will generally consist of some common sense questions about circuit racing and the rules and regulations behind it. This includes items such as flag signals, what to do if you have a mechanical issue, and so on. The test fee varies but is usually between £250 and £300.

You will also be required to have a medical test (carried out by your GP) before you can apply for a licence. Again, there will often be a fee for this.

You have to be at least 16 years old in order to take the test, although a junior test is available for racing in the junior categories. You only have to pass it once, but you will need to renew your MSA licence each year. Once you reach the age of 45, you must have a medical every year.

Once the ARDS test and medical are complete, you can fill out the application form for the National B licence, along with payment of around £55. Your credit card-sized licence will be processed and sent to you.

8.2.2 Race wear and budget

The race series you decide to race in will determine what protective clothing you need and what budget you will require to be competitive.

Budget predictions will need to take into account one-off purchases, such as your race wear, the car and transport, while the consumable side of the budget will need to include fuel, tyres, maintenance, upgrades, damage, entry fees and registration fees (for the racing club and the championship series). The upkeep and running of your transport will also need to be factored in.

Your race wear budget could vary from around £800 for all of the necessary kit, excluding a HANS device (head and neck support), to around £4000 if you decide to go for the best of everything, including a HANS device.

For your crash helmet, you may want a range of visors (e.g. clear, smoked, iridium), with anti-fog spray for the inside of the visor and a rain repellent for the outside of the visor. All race kit should be fireproof and, in most instances, will need to be to a standard of FIA 8856-2000 (see the MSA Blue Book for more information) and Snell SA2005 for the helmet.

crash helmet

HANS device

protective suit

Figure 8.6 *A range of race wear*

8.2.3 Race clubs

There is a huge range of race clubs and championships to suit everyone's budgets and preferences of event type. You can look at a wide range of championship types by looking at the websites of racing clubs, such as the Castle Combe Racing Club (CCRC), British Automobile Racing Club (BARC), British Racing and Sports Car Club (BRSCC), 750 Motor Club (750MC), Motorsport Vision Racing (MSVR), Classic Sports Car Club (CSCC) and Historic Sports Car Club (HSCC).

Once you have decided on the type of racing you want to do, you can find a series that fits your preference, join the club, register for the championship and then find a car.

8.2.4 Other decisions

Other avenues that you may explore when considering racing are covered below.

Sponsorship

This is a great opportunity to gain additional finance, technical support and parts/services discounts. It is not easy to gain sponsorship as you need to find a company that will be able to see a benefit from investing in you and your team and will be able to see a return, although a lot of motorsport-based sponsorship will be an emotional buy-in.

Build a car or buy a ready-built car?

Building a car from scratch will allow you to build the car as you want it to be; however build and development can be very time-consuming and is, of course, expensive. Buying a ready-built front-running car is an easy way to quickly get in to racing as the car should be reliable and developed already and sold at a lower cost than building one yourself.

Buy or hire a car?

Depending on your personal situation with regard to storage, finances and future plans, you may decide to hire a car rather than buying one outright.

Privateer or team?

If you have the facilities and equipment, along with knowledge and skills, you can easily run the car yourself. However, running with a team will generally mean they prepare the car prior to the event, transport it there, and then run it on the event. This allows a more relaxed and focused approach to the event for the driver, while the team members use their expertise to prepare and set the car up.

Driving skills improvement

There are two key ways to improve your driving and reduce lap times. The first method is to track time with data logging – being comfortable in the car and 'at one' with it will provide you with the confidence to push harder and analysing the data will further improve times. To further improve your skills, you can also work with a professional driver coaching expert. Driving simulators are also becoming popular across a wide range of circuit racing series.

8.3 Set-up

Race car set-up is a continual development programme and you must push forward between every event and every season.

In order to produce a race-winning car, you need to find a good baseline set-up for the car to begin with. This should produce a car that is quick to drive around most circuits in the country, and then you can tailor the car and driver to each circuit and its characteristics.

8.3.1 Know your circuits

Each circuit will require a slight adjustment in gearing, aerodynamics and chassis set-up in order to produce the best package. Some circuits will require a similar set-up to each other, while others will be radically different. You need to learn the characteristics of each circuit. The most important thing is to take feedback from the driver and from the car to create a database of information for each circuit; you can build upon your baseline for the event and in future when you return to the circuit. As discussed earlier, it is important to walk around the circuit to take note of the track's individual characteristics, which may not be obvious to the driver when in the car. This can include gradient changes in the circuit, surface changes, kerb aggressiveness, braking point markers and other useful information that could help build the package.

8.3.2 One change at a time

Unless you are sure of what a change is going to do to the car, you should only make one change at a time in order to understand what that change has done. Making multiple changes could potentially mask one change that has actually made the issue worse. Incremental set-up changes allow for them to be measured and also ensure the driver is not caught out by a car whose handling has been dramatically altered.

8.3.3 Set-up elements

Before the event, the car must be set up based on previous data that has been collected, or to the baseline set-up. This initially includes the gear ratios (see Chapter 2 Transmission) and the aerodynamic package (see Chapter 4 Aerodynamics).

The main focus, after the preparation of the car, is the basic chassis set-up, including ride heights, camber, caster, toe and corner weights. Brake bias can also be set after this by simply pressing the brake pedal and getting the front wheels to lock just before the rears in an appropriate area – this ensures that the brake bias is set approximately in the correct position and will then allow the driver to make minor changes on track.

It is important that the chassis set-up is carried out in an accurate, consistent and repeatable manner – this is only achievable with the right equipment and a team that can use it.

The set-up process must be done with the car in a race-ready state. This means it should be full of fluids with half race fuel in the tank (as an average fuel load) and the driver, or similar equivalent weight, in the seat. Bodywork must be removed to allow adjustment of the suspension. Anti-roll bars at the front and rear must be disconnected and dampers set to full soft in order to remove any stiction or unwanted preload – as we are looking at each corner individually, we do not want any interference from any other area of the car.

The wishbones and pushrods should all be checked for damage and the pushrods/damper lengths should be set equally side to side, with ride heights and corner weights set to within 10–15% of target – this helps to see if there are any underlying issues with the car. The wheelbase on either side of the car should also be frequently checked as this can sometimes be out. The chosen spring rates, bump rubbers and packers (these are spacers that determine when the damper body comes into contact with the bump rubber on the damper shaft) should be fitted. Any ballast that will be needed for the race should also be fitted low to the car and as near to the centre of gravity as possible, and as per the regulations' requirements.

8.3.4 The flat patch

Once the car is race ready, it should have its ride height and geometry set up on a flat patch. A flat patch is also known as a flat floor; it is important to have this in order to gain accuracy and consistency for ride heights, caster and

Figure 8.7 Adjustable flat patch and set-up wheels

camber. Toe, however, does not necessarily need to be set on a flat floor.

A flat patch can be either professionally laid concrete, which is great for the workshop floor, or created using a set of adjustable pads, which can then also be transported to and from the event.

When setting the flat floor you must ensure that it is free of any debris and is set on a firm and solid base. The four corner pads need to be set so that there is less than 0.01° in level between the left and right sides; the front and rear pads can be within 0.1°. An engineering level and straightedge can be used to set this. You should also mark the corner pads' feet in relation to the floor, which will help for accurate replacement if the patch is moved from its current location.

With the car on the flat patch, the steering wheel should be locked in its central position with either a steering wheel chock or rack stops (the rack displacement should also be equal for the left and right side). The wheel and tyre combination should be the same as the ones the car will race with and should be set to the target hot tyre pressures, or a set of dummy set-up wheels can be manufactured. Remember, that if you use a set of wheels that have any rim damage or are bent or buckled; your geometry measurements will be inaccurate.

8.3.5 Setting ride height

Ride height should then be set. Note that the minimum ground clearance of the car according to the MSA regulations for most classes of circuit racing is 40 mm (unless otherwise stated in the series regulations); this is the measurement from the lowest part of the car to the floor. Ride height differs because it is a measurement to determine the distance of the chassis to the floor, measured both at the front and rear of a specific point on the car. It should be known or marked so that it can be repeatedly measured from the same point. It is common practice to use chassis rake, in which the front of the car is lower than the rear of the car in order to give an aerodynamic benefit and move the centre of pressure forward on the car. When lowered, the ride height can reduce the centre of gravity height, but be careful not to allow the car to bottom out on the circuit as this will result in a loss of control for the driver (although skid plates are often used to protect areas of the car from touching the surface when low heights are used). Remember that ride height can affect the geometry of the car and its aerodynamic balance and efficiency. Ride height should be set on the pushrods (if available) or the spring platforms of the dampers if conventional sports racer/single-seater suspension is not used.

Ride height can be measured with a range of tools, including steel rules, callipers and laser/ultrasonic sensors; but machined ride height blocks are usually the easiest, quickest, cheapest and most accurate option in my opinion. The ride height should be measured on all four corners of the predetermined 'front' and 'rear' point. This may be a bulkhead section of a specific chassis rail.

While setting ride height you must be aware of the minimum ground clearance rule and this is often used to set the front splitter height (which will be the lowest part of the car). So you may have a car with a 40 mm front splitter height (although regulations may mean that you set this slightly higher due to low fuel loads), which is the lowest part of the car, and then a front ride height of 60 mm and a rear ride height of 68 mm. It is good practice to raise the car off the floor to make ride height alterations as it reduces the load on the threads and avoids damaging them.

When adjusting anything on the car, bounce the car up and down between adjustments in order to settle the suspension and have the four wheels on turn plates to remove any stiction between the tyres and the ground.

8.3.6 Setting the other variables

Geometry

Once the ride height has been set, the geometry will need to be set. Depending on the suspension design and the adjustability, you should set caster and camber before setting toe. Some cars allow for the order to be ignored, which is when shims are used for camber adjustment.

Figure 8.8 Four-wheel alignment equipment

Figure 8.9 String geometry method

Caster

This should then be set and a variety of tools can be used for measuring this. Laser four-wheel alignment sets, such as those supplied by Supertracker along with its camber/caster gauge, are easy to use, quick to set up and provide accurate results. Caster should be set equally on both sides of the car so that it handles consistently through both left- and right-hand turns.

Camber

This should be the next thing to set and, again, there are a range of tools available that can be used to measure it with. Camber often differs across the four corners of the car as each corner will have different loads put through it during an event. Most notably, on a clockwise circuit, you would expect the left front tyre to have the most camber and the right rear to have the least. Remember, as you saw in Chapter 3 Chassis, there is a balance with camber between cornering power and braking/accelerating power.

Toe

Also known as wheel alignment, this is the next variable to adjust on the car, and on some cars has to be left until last because adjusting camber can also alter the toe setting, based on the design of the upright and the wishbone mounting points. Toe is often adjusted with a string and bar method or, if you have the equipment, a laser alignment kit can be used to

Figure 8.10 Camber gauges of various types; digital and spirit level type

set all three parameters. The alignment of the car firstly allows the rear to be aligned so that any thrust angle is removed; this makes the car 'square'. The actual toe in/out measurements can then be made.

8.3.7 Corner weighting

Corner weighting is the final process of the chassis set-up and should be carried out with the car set-up as above. The principle of corner weights is to equalise the diagonal weights of the car so: LF + RR = RF + LR. You must ensure that the scales are zeroed before carrying this out. In order to equalise the weights, you should either increase the ride height of the lighter diagonal or reduce the height of the heavier diagonal. The more symmetrical the car is longitudinally, the better the left to right split will be, so single-seaters should be 50:50, while a saloon car will often be very different.

Once the car is set, you should recheck all measurements as it can sometimes require multiple adjustments. Once done, the anti-roll bars should be refitted and adjusted so that there is no preload applied in order to fit them, the dampers should be reset, driver ballast removed and all parts that have been adjusted should be checked for the correct tightness.

Figure 8.11 Corner weighting

8.3.8 Set-down check

After completing a session or event, you should perform a 'set-down' check to see if anything has moved at all in its set-up, which is always possible if the driver has hit the kerbs too hard, gone off the track or made contact with another car. You should check all the settings already discussed and if the car has an underlying issue in the session, such as independent wheel locking or different handling characteristics during left- and right-hand turns, this may reveal a corner weight issue. You may even find that the car handles better after a clash with the kerbs or another car and that the newly altered set-up should be taken as the new baseline set-up for that circuit!

The adjustment of all of these parameters can be made via the following:

- Shims – these are the best method of adjustment, as they separate both camber and toe adjustment so that they do not interact with each other. Some cars also use them for pushrod length adjustment. The shims have different thicknesses and, with prior measurement, it can be easy to determine camber adjustment and value without any measurements taking place. This is a very good system for switching from dry to wet conditions or vice versa – a reduction or increase in camber.
- Barrel nuts – this set-up consists of a rose joint inserted in a female threaded bush – the barrel (sat in a plain bush, which is part of the wishbone). By loosening the lock nut and winding it one way or the other, you will move the rod end inwards or outwards, giving infinite adjustment to gain the geometric measurement required.
- Rod ends – these can be fitted into the suspension arms and secured with a lock nut. They give good levels of adjustment by partially dismantling the attached items and rotating the join half a turn at a time.
- Suspension/steering arms with a left-hand thread at one end and a right-hand thread at the other. These allow for a lock nut to be loosened at each end of the linkage, and then for the threaded bar to be simply rotated in order to wind the rod end threads in or out at both sides.

rod end

8.3.9 Changes according to weather conditions

Wet weather racing can cause panic among some teams, but as long as you are prepared, it should be no more of an issue than dry weather racing. Ensure that you have the correct equipment and a capable team around you, who are versed in switching from a dry to wet set-up or vice versa.

If conditions change, you will need to consider a method for carrying the wheels and tyres, jacks, wheel guns, torque wrenches and any other associated tools. The driver will need to be protected with a decent umbrella and will need to

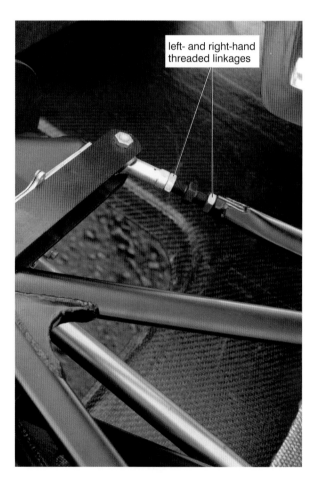

Figure 8.12 Shims, barrel nuts, rod ends and left- and right-hand threaded linkages

ensure that the pedal box areas and their boots before they get into the car are dried properly.

The set-up of the car will need to be altered according to the conditions – this may include aerodynamic changes to increase downforce (dependent upon time and equipment available), suspension changes that include softening of damping and anti-roll bar settings, and taping up the brake and radiator ducts in order to keep everything at an optimum temperature. Although difficult to plan and often avoided by teams, testing should be carried out in both dry and wet conditions in order to explore what settings work for each condition. Just guessing the wet set-up of the car by softening everything is not the best solution. Wet set-ups should also be recorded on a set-up sheet.

Radical Set-up Sheet

Date	
Venue	
Event type	
Session/race number	

Front Nik-link	

Tyre temperatures (°C)		Out	Mid	In
	Hot			
Tyre pressures (°C)	Cold			
	Hot			

Camber (degrees)			negative	
Toe (mm)			out	

Damper setting		B	R	clicks

Ride height		mm
Corner weight		kg
Spring preload		turns
Spring rate		lb

Tyre temperatures (°C)		Out	Mid	In
	Hot			
Tyre pressures (°C)	Cold			
	Hot			

Camber (degrees)			negative	
Toe (mm)			in	

Damper setting		B	R	clicks

Ride height		mm
Corner weight		kg
Spring rate		lb

Rear Nik-link	
Ambient air temperature (°C)	
Track temperature (°C)	

Gear ratio sprockets (F/R)	
Wing position	
Drive planes	

Tyre description	
Weather condition	

Figure 8.13 Set-up/run sheet

motorsport

Tyre temperatures (°C)		In	Mid	Out
	Hot			
Tyre pressures (°C)	Cold			
	Hot			

Camber (degrees)		negative
Toe (mm)		out

Damper setting	B	R	clicks

Ride height		mm
Corner weight		kg
Spring preload		turns
Spring rate		lb

Tyre temperatures (°C)		In	Mid	Out
	Hot			
Tyre pressures (°C)	Cold			
	Hot			

Camber (degrees)		negative
Toe (mm)		in

Damper setting	B	R	clicks

Ride height		mm
Corner weight		kg
Spring rate		lb

Fuel start		Litres
Fuel finish		Litres
Average fuel per lap		Litres

Glossary

A

Acceleration force/g-force: this is caused by a vector sum of non-gravitational forces that acts on the car. The stresses and strains caused by this force are felt as a weight.

Ackerman steering principle: this explains the steering geometry that allows the steered wheels to turn at different rates, when comparing the inner and outer wheel.

Aerodynamics: the study of a fluid (air in this case) with a body (car part or component) moving through it.

Aerodynamic load (downforce): the downward force of air pressure, which increases vertical load on the tyres.

Air/fuel ratio (AFR): a measure of the combustible mixture entering the engine. The ratio determines the mass of air in comparison to one unit of the mass of fuel.

Angle of attack: the angle between the chord line and the freestream air on the wing.

Aspect ratio (of wing): Span ÷ Chord

Attached flow (laminar flow): where airflow follows the surface of the body it is flowing around.

Atomised fuel: fuel in the form of tiny droplets, making it more combustible when mixed with air.

Automatic torque biasing (ATB): type of limited slip differential that is common in motorsport and uses a helical gear set-up to bias torque from one wheel to the other.

B

Belt plies: the chord material used to form the structure of the tyre in order for the rubber to stay attached and in shape.

Billet: a solid piece of material that is then machined into a part with the use of mills, lathes or similar.

Boundary layer: a thin layer of static or slow airflow next to the surface it is flowing over. The air is slow due to surface friction.

Bump steer: the toeing in or out of the wheels when under a bump or droop situation.

C

Calibrated shim: often used for suspension geometry, these specially machined plates of varying thickness determine setting such as toe, camber and caster.

Calliper: used in the braking system to force the brake pads against the disc. The calliper is activated by fluid pressure forcing multiple pistons out to the back face of the brake pad.

Camber line: race car wings need to have camber in order to optimise the car for efficiency. The camber means that the lower surface is more curved than the upper surface. Therefore, the camber line is always an equal distance from the upper and lower surfaces from the leading to trailing edge.

Controller area network (CAN) bus link: an electronic mode of communication for control units around a vehicle.

Cast: a machining process carried out by filling a mould with liquid hot metal and allowing it to cool in the mould to form a shape. Engine casings are commonly cast.

Caster (degrees): the angle of the steering axis compared to a vertical line through the centre of the wheel when viewed from the side. Positive caster is used to provide an increase in negative camber when steering.

Cavitation: the formation of air bubbles caused by fluids being worked and accelerated quickly, resulting in frothing. This affects the working properties of the fluid.

Centre of gravity (CG): an imaginary point in which all forces act through on the car and is effectively the pivoting point at which the car would remain balanced.

Centre of pressure: the point at which the aerodynamic forces are balanced. Ideally this should be in close proximity to the centre of gravity in order to provide a stable car in high- and low-speed scenarios.

Chord: the distance from leading to trailing edge.

Coefficient of drag (Cd): this is a unitless number that allows comparison of drag incurred by different-sized and shaped bodies.

Coefficient of lift (Cl): in this case downforce, this is a unitless number that allows comparison of downforce (negative lift) incurred by different-sized and shaped bodies.

Computational fluid dynamics (CFD) testing: a theoretical way of wind tunnel testing using a high-powered computer and intelligent software package.

Conrod: the link between a piston and crankshaft which assists with converting the reciprocating motion into a rotational motion.

Contact patch: the surface area of the tyre that is in contact with the road surface.

Crabbing: caused by poor wheel alignment, when the rear wheels try to steer the car rather than tracking behind the front wheels.

Crankshaft: this component is connected to the piston via a conrod, providing a way to operate the drivetrain.

Crown wheel: the large geared wheel in a differential that can also determine the final drive ratio.

D

Damper shaft: the part of the damper to which the piston (inside the damper body) and the top mounting eyelet is connected.

Damping rate: the amount of energy-absorbing power that a damper has to control the spring's motion.

Dive plane: an aerodynamic trimming device, taking the form of a curved plate that points upwards at the rear to add downforce.

Dog engagement: a way of engaging the meshed selector ring with the gear required.

Doppio corpo orizzontale (DCOE): double throat horizontal die cast carburettor.

Downforce: see 'aerodynamic load'.

Drag: the longitudinal force that faces the car head-on and reduces horsepower at any velocity – the higher the velocity, the more horsepower is needed to overcome drag.

Driveshaft (or half shaft): a hollow or solid bar, with splines at either end, used to connect the differential with the hubs.

Droop travel: the amount of rebound travel (the wheel moving down) in the suspension.

Dynamic pressure: movement energy of the air (or fluid) expressed as $0.5 \times \rho \times V2$.

Dynamometer (dyno): a computer-controlled machine that is used to test equipment. Most commonly in motorsport are engine dynos, damper dynos and chassis test rig dynos.

E

Electronic control unit (ECU): this is most commonly associated with the engine system to control ignition and fuelling.

End plate: flat sheet fitted to the end of a wing profile to reduce spillage of high-pressure air on the upper surface to the low-pressure region underneath.

Engine management system (EMS): most commonly an electronic system that consists of an electronic control unit, loom, sensors and output devices to time and control fuel input and spark.

Excitation phase: when an electronic component is supplied with a voltage to perform an operation.

F

Flat patch: a completely flat surface used to carry out chassis set-up work.

Flywheel: this rotating disc is the direct link between the crankshaft and clutch.

Forge: a manufacturing process that relies on a malleable material to be heated, pressed and worked into shape.

Freestream: airflow around the body which is undisturbed by the body moving through it.

Full cylinders: see 'volumetric efficiency'.

Full droop: the point of maximum suspension travel in the wheel's downward position.

G

Green flag lap: a lap at the start of the race to allow tyres, brakes and engine to be brought up to temperature.

Green track: a slippery surface condition where rain causes contaminants to rise to the surface of the track, leaving it slippery when dry.

Ground effect: airflow underneath the body that is modified by the body running in close proximity to it.

Gudgeon pin: a tube that connects the connecting rod to the piston.

Gurney flap: right-angled (in most cases) strip that is most commonly fitted to the rearmost wing element to add additional downforce.

H

Half shaft: see 'driveshaft'.

Hall effect sensor: transducer that varies its output voltage in response to a magnetic field.

HANS device: head and neck support worn by drivers to limit the amount of head movement available, protecting the neck during an impact.

Heat cycle: when the tyres go through a heating and then a cooling phase.

Helical gear: a type of gear cut used on road cars where the teeth are cut at an angle so full engagement is gradual.

High-density air: air that contains a high number of closely packed molecules and, therefore, a high amount of oxygen.

Hydroscopic fluid: fluid that absorbs water.

I

Induced drag: drag caused directly by vehicle body shape and layout.

Inertia: the resistance of any physical object to a change in its state of motion.

Inlet manifold: engine component that joins the carburettors or throttle bodies to the cylinder head.

Inlet runner: length of the inlet measured from the back of the inlet valve to the opening of the inlet trumpets.

Input shaft/clutch shaft: machined shaft in the gearbox that connects to the engine via the clutch and is used to hold the driver gears.

Instantaneous centre (IC) point: the point at which geometric lines meet each other.

L

Laminar flow: see 'attached flow'.

Lap beacon: a device used on the pit wall that acts as a trigger for a lap timer to show the driver lap times.

Lay shaft/output shaft: shaft that holds the driven gears and connects to the input shaft gears and transfers this movement to the crown wheel of the differential.

Leading edge: front tip of the wing.

Lock wire: twisted and tensioned wire that links a fastener to a secure object to prevent it from coming loose.

M

MSA Blue Book: the competitors' and officials' rules and regulations book for most UK motorsport events.

P

Parc fermé: a secure area where the race cars and drivers are directed to after the race has finished. It is used to carry out eligibility checks by the officials.

Peak grip: the limit at which tyre grip is at its best, often measured through g-force and slip angle.

Pitch: the action of the chassis pivoting along a transverse axis from acceleration and braking forces.

Pitot tube: a method of measuring air speed by comparing dynamic and static wind speeds.

Pivot point arch louvers: vents placed strategically above the wheels in order to release the high pressure in the wheel arches and reduce lift.

Planet gears: type of gears used within a differential.

Potentiometer (pot): a sensor used for positional measurements.

Power: how much work can be done in a given time; it is a function of torque and engine speed.

Powerband: the part of the engine's rev range where optimal power is produced.

Preload: the compressive force applied to the spring when fitted to the a damper, prior to being loaded up by the car's mass.

Propshaft: the rotating shaft that links the gearbox output to the differential input.

Push/pull rod system: a method of activating the dampers, linking the wheels through a push/pull rod and a bell crank connected to the damper.

R

Random-access memory (RAM): a form of computer data storage that allows all data to be accessed in the same amount of time.

Ride frequency: the undamped natural frequency of the body in ride.

Ride height: a set of points used to measure the attitude of the chassis.

Ride rate: the way the car tackles bumps and undulations under normal straight-line conditions.

Roll: when the car experiences a lateral g-force, the chassis will roll towards the outside of the corner; the amount depends upon roll stiffness levels.

Roll axis: a line linking the front roll centre to the rear roll centre.

Roll centre: the point that the sprung mass rotates around.

Roll moment: the inertia of the sprung mass of the car during roll.

Rotary variable differential transducer (RVDT): an electromechanical transducer, which provides an alternating current (AC) output that is linearly proportional to the angular movement of the input shaft.

Read-only memory (ROM): a form of computer data storage where the data cannot be changed.

Power (hp): (Engine rpm × Torque (lb/ft)) ÷ 5252

Rubbered in track: where the weather conditions allow the circuit to be cleaned and rubber is laid down over the tarmac by other cars, providing additional grip (the opposite of 'green track').

S

Scavenging exhaust: a tuned exhaust system that helps pull burnt gases out of the engine and draw in fresh mixture during valve overlap.

Separated flow: airflow that no longer follows the shape of the body.

Sequential gear layout: replacing the conventional H-gate layout, this requires a simple forward–backwards motion of the gearstick, provided by the addition of a rotating barrel within the gearbox with individually machined pathways for each selector fork.

Shock body: the outer section of the damper, which houses all the components and fluid.

Side bevel gears: conically shaped gears.

Side skirt: horizontal extension that protrudes from the side pods, between the front and rear wheels. It is often used to help protect the low-pressure underfloor of the vehicle from the high-pressure area around the outside of the car body.

Slip angle: the angle between the wheel's rolling direction and the way it is actually pointing.

Span: the width of the wing from one end to the other.

Spline: a method for keying two components together. Splines are often rolled onto a bar to allow the 'teeth' to fix onto another part. This is often used for driveshafts and gearbox shafts.

Splitter: a generally flat, horizontal-forward extension of the bodywork that acts as a cutting device to separate the air going over and under the car. The splitter works by acting as an extension of the floor so that a low pressure can be created underneath it, while the upper surface acts as a shelf to harness the high-pressure air sitting around the front of the bodywork.

Spring rate: the amount of force required to compress a spring a specific distance.

Sprung mass: the weight of the vehicle that is supported by its suspension springs, including the chassis, engine, driver and gearbox. The heavier the sprung mass, the stiffer the spring must be to keep the car off of the ground.

Stagnation point: the point, usually at the front of the body, where the air velocity is zero and air pressure is high.

Static pressure: the ambient pressure present within a certain area.

Stiction: this is the static friction caused by friction between bearing and pivot point faces, as well as the tyre and floor surface. This should be known and taken into account when carrying out chassis set-up.

Strakes: these are used to assist airflow and can often be seen in diffuser tunnels.

Stressed member: a section of the car that is used to mount another system on, so will have additional load passed through it other than its main form of operation (e.g. the rear suspension mounted on the gearbox).

Surface drag: drag caused by the friction of the body's surface.

Synchromesh: a system that engages the selector ring with the desired gear, often replaced for a dog engagement type in motorsport.

T

Thermal efficiency: the measure of how heat energy is used. In an engine, the thermal efficiency determines how much of the heat energy is used to push the piston back down the cylinder.

Thermistor sensor: a temperature sensor that changes electrical resistance, and therefore voltage output, in relation to its temperature. It can be either a negative temperature coefficient (NTC) or positive temperature coefficient (PTC) type.

Thermocouple sensor: a temperature sensor made up of two wires of different materials welded together into a junction. A change in temperature within this measurement junction generates a current in the wires proportional to the temperature change.

Throttle position sensor (TPS): this is used for datalogging and engine management purposes.

Thrust angle: the angle that the rear wheels are driving towards. Any thrust angle should be removed during the alignment process.

Top dead centre (TDC): the position when the piston has reached its highest position in the cylinder.

Torque: a measure of the force that is applied on a lever, multiplied by the distance to the rotation point. Torque = Force × Distance from pivot.

Traction circle: this is the same as a g-plot, which looks at longitudinal g-force vs. lateral g-force to determine the performance of the car, tyres and driver.

Trailing edge: rear tip of the wing.

Turbulent flow: flow that begins to swirl or mix while flowing around the body.

U

Unsprung mass: the weight that is not supported by the springs, such as wheels, tyres, brakes and uprights. As this is effectively uncontrolled, the lighter it is, the better the contact between the tyre and road surface, particularly during transitions and when tackling bumps and undulations.

V

Valve train: the assembly that controls the opening and closing of the inlet and exhaust valves, including the follower, springs, camshafts, and so on.

Venturi: the venturi effect is when a fluid is passed through a constricted pipe and, as a result, the speed increases and fluid pressure reduces.

Volumetric efficiency: how much air and fuel can be fitted into a cylinder – this figure can be over 100 per cent with some designs.

W

Weight transfer: the movement of weight around the four tyres of the car during a dynamic state. Weight transfer is affected by overall weight, track, wheelbase, centre of gravity and g-force.

Wheel rate: the effective spring rate of the wheel – what the driver feels and the tyres deal with while driving.

Wheel vent: an open area behind the wheels whereby the air pressure can be extracted from the wheel arches to help reduce lift.

Windage tray: similar to a baffled sump, this insulates the crankshaft from the windage phenomenon, where tiny droplets of oil become airborne in the crankcase at high rpm.

Wing tip vortices: this is air that begins to rotate as it spills off of the sides of the wings, unless controlled. It can cause heavy amounts of drag and reduce efficiency.

Y

Yaw angle: the direction the car is facing compared to the direction it is actually moving in.

Index

A

Absorbent glass matt (AGM), batteries 158
Acceleration force/g-force 14, 29, 78, 95, 151, 176, 180
Acceleration sensors 151–152
    piezoelectric accelerometers 151–152
Ackerman steering 104–105
Active wheel speed sensors 153
Aerodynamic load see Downforce
Aerodynamics 125–144
    air density 128–129
    arch louvers 142
    barge boards 143–144
    Bernoulli's theory 126–128
    body shape 128–129, 132–133
    boundary layer 131
    centre of pressure 126
    computational fluid dynamics (CFD) 130
    data logging 129
    diffusers 138–139
    dirty air 132
    dive planes 141
    downforce 126–129, 131–132, 134–138, 139
    drag 126–128, 131–132
    dynamic pressure 131, 154
    factors affecting 131–132
    flow-viz paint 130
    freestream 132
    ground effects 138–139
    induced drag 131–132
    measuring 129–130
    pivot point arch louvers 142
    principles 126–132
    ride height 132–133
    side skirts 142–143
    splitter 140–141
    stagnation point 131
    strakes 143
    surface drag 131–132
    tow 132
    track testing 130
    turbulent flow 132
    turning vanes 143–144
    underfloor 138–139
    vortex generators 143–144
    wheel vents 142
    wind tunnel testing 129
    wings 126–128, 134–138
AFR see Air/fuel ratio
AGM see Absorbent glass matt
Air density, aerodynamics 128–129
Air/fuel ratio (AFR), sensing 154–156
Ampere hour (A/h) rating, batteries 158
Angle of attack (AoA), aerodynamic wings 134
Anodising, protective surface finishes 194
Anti-pitch geometry, suspension 95–96
Anti-roll bars (ARBs), suspension 92
ARBs see Anti-roll bars
Arch louvers, aerodynamics 142
ARDS test 206
Aspect ratio, aerodynamic wings 134
ATB see Automatic torque biasing (ATB) limited-slip differential
Atomised fuel 3–4
Attached flow (laminar flow), aerodynamic wings 134–135
Automatic torque biasing (ATB) limited-slip differential 65–74

B

Baffled wet sump, lubrication 29–30
Barge boards, aerodynamics 143–144
Batteries 158–161
    absorbent glass matt (AGM) 158
    ampere hour (A/h) rating 158
    battery isolators 160–161
    cold crank ampere (CCA) rating 159
    isolators/master switches 160–161
    pulse current rating 159
Belt plies, tyres 82
Bernoulli's theory, aerodynamics 126–128
Billet 26–27, 28, 42, 82
Body shape, aerodynamics 128–129, 132–133
Bolts and nuts, preparing a car 186–191
Bottom end 26–28
Boundary layer, aerodynamics 131
Brake-type rolling road 36
Brakes/braking 105–124
    brake lines 121
    braking basics 105
    calculations 124

Index

callipers 114–118
cooling 121
data logging 121, 178–179
discs 119–120, 121
fluid 121
master cylinders (MCs) 105–114
pad knockback 121
pads 120–121
paint, brake disc temperature 121
pedal box 105–114
weight transfer 107, 114
Budget 206
Bump rubbers, suspension 88
Bump steer, steering 103

C

Calibrated shim 90
Callipers, brake 114–118
Camber
set-up 211, 212
suspension 92
Camber line, aerodynamic wings 134
Camshaft and valve train 18–21
CAN see Controller area network
Car buying/hiring/building 208
Carburettors 7–9
doppio corpo orizzontale (DCOE) 7–8
Cast 26–27, 28, 42, 82
Caster
set-up 211
suspension 93
Cavitation 12, 34, 90
CCA rating see Cold crank ampere (CCA) rating
Centre of gravity (CG), suspension 95
Centre of pressure, aerodynamics 126
CFD see Computational fluid dynamics
CG see Centre of gravity
Changes, set-up 208
Channel reports, data logging 175–176
Chassis analysis, data logging 171, 180–182
Checklists, preparing a car 195–197
Chord, aerodynamic wings 134
Circuit knowledge, set-up 208
Clothing, protective 206–207
Clubs, race 207
Clutch 42–50
friction (or clutch) plate 43–49
pressure plate 50
torque capacity 43

Clutch/plate differential 65–74
Coefficient of drag (Cd) 127
Coefficient of lift (Cl) 127
Coil spring types, suspension springs 83–84
Cold crank ampere (CCA) rating, batteries 159
Combustion chamber 21–22
Component management, preparing a car 193–194
Computational fluid dynamics (CFD), aerodynamics 130
Conrod 4, 26–28
Contact patch, tyres 76–77, 80, 82, 92, 95
Controller area network (CAN) 166, 168
Coolant
engine 33–34
gearbox 62
Cooling, brakes 121
Corner weighting, set-up 212–213
Crabbing 93
Crankshaft 26–28, 42, 153
Cross-ply tyres 76
Crownwheel 65, 74
Cylinder head 16–22
ancillaries 21–22
camshaft and valve train 18–21
exhaust valve and port 18
inlet valve and port 16–17

D

Damper shaft 88, 209
Dampers, suspension 88–91
Damping rate 77, 78
Dash display units 162–166
Data logging 167–184
see also Sensing
aerodynamics 129
brakes/braking 121, 178–179
channel reports 175–176
channels 170
chassis analysis 171, 180–182
components 168–169
downforce 181–182
driver data analysis 171, 178–179
engine analysis 171, 182–183
engine management system (EMS) 146, 168
global positioning system (GPS) sensors 156, 168
histograms 176
lap beacon 168

logging/sample rates 169–170
maths channels 183–184
memory 170
sector times 172–175
software 171–177
systems 168–170
telemetry 170–171
time slip 172–175
track maps 171–175
XY graphs 176–177
DCOE see Doppio corpo orizzontale
Differential 64–74
automatic torque biasing (ATB) LSD 65–74
clutch/plate LSD 65–74
limited-slip differential (LSD) 65–74
locked 65
open 64
Diffusers, aerodynamics 138–139
Dirty air, aerodynamics 132
Discs, brake 119–120, 121
Displacement sensors 149–150
linear variable differential transducers (LVDTs) 149–150
rotary variable differential transducers (RVDTs) 149–150
Dive planes, aerodynamics 141
Dog engagement, gearbox 53–54
Doppio corpo orizzontale (DCOE), carburettors 7–8
Downforce 77, 87
aerodynamics 126–129, 131–132, 134–138, 139
data logging 181–182
Drag
aerodynamics 126–128, 131–132
induced drag 131
surface drag 131
Drive shaft (half shaft) 62, 65
Driver data analysis, data logging 171, 178–179
Driver interface 162–166
Droop travel 86, 95
Dry sump, lubrication 31–32
Dynamic pressure, aerodynamics 131, 154
Dynos
engine 36–37
rolling road 36
tuning 36–37

E

Electrical 146–166
batteries 158–161
battery isolators 160–161
controller area network (CAN) 166, 168
driver interface 162–166
fire extinguishers, electronic 161
rain lights/timing transponders 159–160
sensing 146–156
wiring harnesses 157–158
Electronic fire extinguishers 162
EMS see Engine management system
End plates, aerodynamic wings 136–138
Engine analysis, data logging 171, 182–183
Engine management system (EMS) 146, 168
see also data logging
Events 200–205
housekeeping 205
MSA Blue Book 200
noise 201
organisation 205
planning 201–203
race day 204
spares 200–202, 205
teamwork 205
testing 201
timekeeping 205
Excitation phase 150
Exhaust system 23–26
Exhaust valve and port, cylinder head 18

F

Farringdon SWIS10 steering wheel-mounted display 163, 166
Fasteners
nuts and bolts 186–191
preparing a car 186–193
Fire extinguishers, electronic 162
Flat patch, set-up 209–210
Flow-viz paint, aerodynamics 130
Flywheel 40–42
inertia calculation 40
types 41–42
Forced induction 21–22, 34–35
Forge 26–28
Four-stroke engines 2–3
Freestream, aerodynamics 132
Friction (or clutch) plate 43–49

Fuel injection 9–14
Full cylinders see Volumetric efficiency
Full droop 84

G

Gearbox 50–62
    actuation type 55–56
    coolant 62
    dog engagement 53–54
    engagement type 53–54
    gear casings 62
    gear ratios 56–61
    gear selection 54
    gear types 52
    H-gate 54, 55
    layout 51–52
    lubrication 62
    options 51
    paddle shift 56
    sequential system 54–55
    spline 51–52, 53
    stressed member 62
    synchromesh engagement 53
    torque limitation 61
Geometry
    anti-pitch geometry 95–96
    set-up 210–211
    suspension 92–96
Global positioning system (GPS) sensors 156, 168
Green flag lap 62, 80, 204
Green track 79
Ground effects, aerodynamics 138–139
Gudgeon pin 27
Gurney flaps, aerodynamic wings 136

H

H-gate gearbox 54, 55
Half shaft see Drive shaft (half shaft)
Hall effect sensor 150, 153
HANS device 206, 207
Heat cycle, tyres 81
Heat management, preparing a car 193
Helical gear 51, 52–53, 65
High-density air 3, 21–22, 131
Histograms, data logging 176
Housekeeping, events 205
Hygroscopic fluid 121

I

IC point see Instantaneous centre (IC) point
Ignition system 15–16
Induced drag, aerodynamics 131
Inductive sensors, speed sensors 152–153
Inertia 40–42
    calculation, flywheel 40
Inertia dyno 36
Infrared (IR) sensors, temperature sensors 147, 148
Inlet manifold 6, 7, 12
Inlet runner 6, 7
Inlet system 5–7
    modified system 5–6
    race system set-up 6–7
Inlet valve and port, cylinder head 16–17
Input shaft/clutch shaft 51, 56
Instantaneous centre (IC) point 94, 104
Intermediate tyres 77
IR sensors see Infrared (IR) sensors

L

Lambda sensors 154–156
Laminar flow see Attached flow (laminar flow)
Lap beacon, data logging 168
Laser distance sensors 154
Lay shaft/output shaft 51–52, 53, 56, 63, 65
Leading edge, aerodynamic wings 134
Life charts, preparing a car 193
Limited-slip differential (LSD) 65–74
    automatic torque biasing (ATB) 65–74
    clutch/plate 65–74
Linear variable differential transducers (LVDTs), displacement sensors 149–150
Lock wire, preparing a car 187–188
Locked differential 65
LSD see Limited-slip differential
Lubrication 29–32
    baffled wet sump 29–30
    dry sump 31–32
    gearbox 62
    oil accumulator 30–31
    swinging sump 30
    wet sump 29
LVDTs see Linear variable differential transducers

M

Master cylinders (MCs), brakes 105–114
Master switches/battery isolators 160–161
Maths channels, data logging 183–184
MCs see Master cylinders
Monoshock system, suspension 99–101
MSA Blue Book 159, 160, 200, 201, 206

N

Noise, events 201
Nuts and bolts, preparing a car 186–191

O

Oil accumulator, lubrication 30–31
Open differential 64
Organisation, events 205
Output shaft see Lay shaft/output shaft
Oxygen sensors 154–156

P

Pad knockback, brakes 121
Paddle shift, gearbox 56
Pads, brake 120–121
Paint
    brake disc temperature 121
    flow-viz paint 130
    protective surface finishes 194
Parc fermé 204, 205
PDMs see Power delivery/management
    modules (PDMs or PMMs)
Peak grip, tyres 76, 77, 78
Pedal box 105–114
Piezoelectric accelerometers 151–152
Pitch 107
    anti-pitch geometry 95–96
Pitot tubes 154
Pivot point arch louvers, aerodynamics 142
Planet gears 74
Planning
    events 201–203
    test sessions 201–203
Plating, protective surface finishes 194
PMMs see Power delivery/management
    modules (PDMs or PMMs)
Potentiometer (pot) 149, 169–170, 181
Power 4, 5
Power delivery/management modules (PDMs
    or PMMs) 163–166

Powerband 56–57
Preload
    differential 74
    suspension springs 85–86
Preparing a car 186–197
    checklists 195–197
    component management 193–194
    fasteners 186–193
    heat management 193
    life charts 193
    lock wire 187–188
    nuts and bolts 186–191
    protective surface finishes 194
    thread lock 187
    Torque Seal 186–187
    track tasks 197
    workshop preparation 197
Pressure plate, clutch 50
Pressure sensors 148–149
Propshaft 63
Protective clothing 206–207
Protective surface finishes
    anodising 194
    painting 194
    plating 194
    preparing a car 194
Pulse current rating, batteries 159
Push/pull rod system, suspension 98–99

Q

Qualifying, race day 204

R

Race clubs 207
Race day 204
    qualifying 204
    scrutineering 204
Radial-ply tyres 76
Rain lights/timing transponders 159–160
Read-only memory (ROM) 9
Recording, set-up 216–217
Resistive temperature devices (RTDs),
    temperature sensors 147, 148
Ride frequency 87
Ride height
    aerodynamics 132–133
    set-up 210
Ride rate 83–84, 87
Roll axis, suspension 94

Roll centre, suspension 93–94
Roll moment, suspension 95
Rolling road 36
    brake-type 36
    inertia dyno 36
ROM see Read-only memory
Rotary variable differential transducers
    (RVDTs), displacement sensors 149–150
RTDs see Resistive temperature devices
Rubbered in track 79
RVDTs see Rotary variable differential
    transducers

S

Scavenging exhaust 3–4, 18, 21, 23–24
Scrutineering, race day 204
Sector times, data logging 172–175
Sensing 146–156
    see also Data logging
    acceleration sensors 151–152
    air/fuel ratio (AFR) 154–156
    displacement sensors 149–150
    engine management system (EMS) 146, 168
    global positioning system (GPS) sensors 156
    lambda sensors 154–156
    laser distance sensors 154
    oxygen sensors 154–156
    pitot tubes 154
    pressure sensors 148–149
    speed sensors 152–153
    strain gauges 153–154
    temperature sensors 147–148
Separated flow, aerodynamic wings 134–135
Sequential system gearbox 54–55
Set-down check, set-up 213–214, 215
Set-up 208–217
    camber 211, 212
    caster 211
    changes 208
    circuit knowledge 208
    corner weighting 212–213
    elements 209
    flat patch 209–210
    geometry 210–211
    recording 216–217
    ride height 210
    set-down check 213–214, 215
    set-up/run sheets 216–217
    toe 211–212

    weather conditions 214–216
    wheels 210–212
Shock body 91
Side bevel gears 74
Side skirts, aerodynamics 142–143
Skills improvement 208
Slick tyres 77
Slip angle, tyres 77–78
Software
    data logging 171–177
    suspension 97–98
Span, aerodynamic wings 134
Spares, events 200–202, 205
Speed sensors 152–153
    active wheel speed sensors 153
    inductive sensors 152–153
Spline
    gearbox 51–52, 53
    steering 102
Splitter, aerodynamics 140–141
Sponsorship 207–208
Spring/damper activation linkage systems,
    suspension 98–101
Spring rates, suspension 87
Springs, suspension 83–88
    coil spring types 83–84
    motion ratio 87–88
    preload 85–86
    specifications 84–85
    spring rates 87
    torsion bar springs 86–87
Sprung mass, suspension 83
Stagnation point, aerodynamics 131
Static pressure, aerodynamic wings 136
Steering 102–105
    Ackerman 104–105
    bump steer 103
    ratios 102
    spline 102
Stiction 186, 209, 210
Strain gauges 153–154
Strakes, aerodynamics 143
Stressed member, gearbox 62
Supercharger 21–22, 34–35
Surface drag, aerodynamics 131
Suspension 82–101
    anti-pitch geometry 95–96
    anti-roll bars (ARBs) 92
    bump rubbers 88

camber 92
caster 93
centre of gravity (CG) 95
dampers 88–91
geometry 92–96
monoshock system 99–101
push/pull rod system 98–99
roll axis 94
roll centre 93–94
roll moment 95
software 97–98
spring/damper activation linkage systems 98–101
springs 83–88
sprung mass 83
third spring set-up 99, 101
toe 92–93
unsprung mass 83
weight transfer 95–96, 107
wishbone layout 96–98
Swinging sump, lubrication 30
Synchromesh engagement, gearbox 53

T

TDC see Top dead centre
Team/privateer 208
Teamwork, events 205
Telemetry, data logging 170–171
Temperature sensors 147–148
    infrared (IR) sensors 147, 148
    resistive temperature devices (RTDs) 147, 148
    thermistors 148
    thermocouples 147, 148
Test sessions, events 201–203
Testing, events 201
Thermal efficiency 22, 33
Thermistors, temperature sensors 148
Thermocouples, temperature sensors 147, 148
Third spring set-up, suspension 99, 101
Thread lock, preparing a car 187
Throttle position sensor 10, 171
Thrust angle 93, 212
Time slip, data logging 172–175
Timekeeping, events 205
Timing transponders/rain lights 159–160
Toe
    set-up 211–212
    suspension 92–93

Top dead centre (TDC) 2, 3, 15–16, 18–19, 22, 151
Torque 4, 5
Torque capacity, clutch 43
Torque limitation, gearbox 61
Torque Seal, preparing a car 186–187
Torsion bar springs, suspension springs 86–87
Tow, aerodynamic 132
Track maps, data logging 171–175
Track tasks, preparing a car 197
Track testing, aerodynamics 130
Traction circle 78
Trailing edge, aerodynamic wings 134
Transponders/rain lights 159–160
Tuning 3–5
    dynos 36–37
Turbocharger 21–22, 34–35
Turbulent flow, aerodynamics 132
Turning vanes, aerodynamics 143–144
Twisted wing elements, aerodynamic wings 136
Two-stroke engines 2
Tyres 76–82
    bedding in 82
    belt plies 82
    blistering 81
    cleaning 82
    construction 76–78
    contact patch 76–77, 80, 82, 92, 95
    cross-ply 76
    degradation 80–82
    graining 81
    hardness 81
    heat cycle 81
    intermediate 77
    peak grip 76, 77, 78
    performance maximisation 77–80
    pressure 79
    radial-ply 76
    slick 77
    slip angle 77–78
    storage 82
    structure 76–78
    temperature 80
    tread depth 82
    vertical load 77–78
    wet 77

U

Underfloor aerodynamics 138–139
Unsprung mass, suspension 83

V

Valve train 16, 18–21
Venturi 7, 9, 139
Volumetric efficiency 4, 6–7, 21–22, 23, 34
Vortex generators, aerodynamics 143–144
Vortices, wingtip, aerodynamic wings 136–137

W

Weather conditions, set-up 214–216
Weight transfer
    brakes/braking 107, 114
    suspension 95–96, 107
Wet sump, lubrication 29
Wet tyres 77
Wheel rate 87–88, 97
Wheel vents, aerodynamics 142
Wheels 82
    set-up 210–212
Wind tunnel testing, aerodynamics 129
Windage tray 32
Wing tip vortices, aerodynamic wings 136–137

Wings, aerodynamic 126–128, 134–138
    adjustment 137–138
    angle of attack (AoA) 134
    aspect ratio 134
    attached flow (laminar flow) 134–135
    camber line 134
    chord 134
    end plates 136–138
    gurney flaps 136
    leading edge 134
    separated flow 134–135
    span 134
    static pressure 136
    trailing edge 134
    twisted wing elements 136
    wing tip vortices 136–137
Wiring harnesses 157–158
Wishbone suspension 96–98
Workshop preparation 197

X

XY graphs, data logging 176–177

Y

Yaw angle 83, 129, 133, 171

Acknowledgements

The author and the publisher would also like to thank the following for permission to reproduce material:

Introduction
Page iii Josh Smith; 0.1 Keith Lowes; 0.2 Jamie Peter Ennis; 0.3 Jamey Price/LAT Photographic.

Chapter 1
1 Radical Performance Engines Ltd; 1.1 and 1.2 LJ Create; 1.3 Anderson Racing Engines; 1.4 Jenvey Dynamics; 1.5 QED Motorsport Ltd qedmotorsport.co.uk; 1.6 Anderson Racing Engines; 1.7 and 1.8 Kinsler Fuel Injection; 1.10 QED Motorsport Ltd qedmotorsport.co.uk; 1.12 Anderson Racing Engines; 1.15 QED Motorsport Ltd qedmotorsport.co.uk; 1.18 Double S Exhausts Ltd; 1.19 QED Motorsport Ltd qedmotorsport.co.uk; 1.20 LUNATI; 1.23 Extreme Engines; 1.27 and 1.29 Radical Performance Engines; 1.30 Anderson Racing Engines; 1.31 Dynojet Research, Inc.

Chapter 2
2 Pro-Shift Technologies Ltd; 2.1 Exedy (L) and Anderson Racing Engines (R); 2.2 Anderson Racing Engines; 2.3 Exedy; 2.4 Zoran Simin/Alamy (Sprung) and Fotolia/Ruslan Kudrin (Unsprung); 2.5 Exedy (organic and three-paddle) AP Racing (four-paddle); 2.6 and 2.7 AP Racing; 2.8 ACT Advanced Clutch Technology; 2.9 popova48/Fotolia; 2.11 Fikmik/Alamy; 2.12 ugo ambroggio/Alamy; 2.17 Curtis Jacobson; 2.18 Firman West Cars; 2.19 Hewland; 2.20 Xtrac; 2.23 Radical Sportcars Ltd; 2.24 Rob Collingridge; 2.26 and 2.27 Quaife; 2.28, 2.29 and 2.30 Xtrac; 2.31 Hewland.

Quaife also supported us with technical editorial input to the differential section, which we would like to thank and acknowledge them for.

Chapter 3
3 Chiron World Sports Cars Ltd; 3.6 Ionut Pascut; 3.7 Sutton Motorsport Images; 3.10 Josh Smith; 3.12 H&R Special Springs; 3.13 Josh Smith; 3.14 Longacre Racing Products Inc http://www.longacreracing.com; 3.20 Koni; 3.22 hrpworld; 3.24 Josh Smith; 3.32 Performance trends; 3.34, 3.35, 3.36, 3.37 Josh Smith; 3.40 Competition Supplies Ltd; 3.41 © 2012 Wilwood Engineering, Inc. - All Rights Reserved; 3.42 Josh Smith; 3.43, 3.44 and 3.45 AP Racing; 3.46 Josh Smith (L) Copyright © 2012 Wilwood Engineering, Inc. - All Rights Reserved (R); 3.47 Copyright © 2012 Wilwood Engineering, Inc. - All Rights Reserved; 3.48 and 3.49 A P Racing; 3.50 Copyright © 2012 Wilwood Engineering, Inc. - All Rights Reserved; 3.51 Chiron World Sports Cars Ltd; 3.54 A P Racing.

Chapter 4
4 Courtesy of ANSYS, Inc; 4.3 Terry Smith; 4.4 mst7022; 4.5 MIRA Ltd; 4.6 Takeshi Takahara/Science Photo Library; 4.7 Courtesy of ANSYS, Inc; 4.8 and 4.9 Sutton Motorsport Images; 4.11 1985 image- Noel Yates/Alamy, 2012 image- Ferrari Press Office/DPA/Press Association Images- ordered; 4.15 Seibon Carbon; 4.17 Josh Smith; 4.18a Christopher Kelly/ActionPlus; 4.18b RFRCARS US; 4.19a Rennie Clayton Dauntless Racing Cars; 4.19b idp oulton park collection/Alamy; 4.21 Radical Sports Cars; 4.22 Steve Nesius/AP/Press Association Images- ordered; 4.23 Red Bull Racing Ltd; 4.24 Josh Smith; 4.25 Michael J. Fuller, www.mulsannescorner.com; 4.26, 4.27 and 4.28 Josh Smith; 4.29 Sutton Motorsport Images.

Chapter 5
5 Josh Smith; 5.1a Pico Technology (T); 5.1b Auber Instruments (M); 5.1c Lorenzo Bellanca/LAT Photographic (B); 5.2a Ruslan Kudrin/Alamy; 5.3a Penny and Giles (TL); 5.3b Penny and Giles (TR); 5.3c Penny and Giles (BL); 5.3e Race Technology Ltd. (BR); 5.7 Josh Smith; 5.8 David J. Green – technology/Alamy; 5.9 Rex Features; 5.13 Josh Smith; 5.14 Tim Mason, TMME Electronics; 5.15 Josh Smith; 5.19 Tim Mason, TMME Electronics; 5.21 Farringdon Instruments.

Chapter 6
6, 6.2, 6.3, 6.4, 6.5 and 6.6 Race Technology Ltd; 6.7 Polylogic Ltd. (T) and Race Technology Ltd. (B); 6.8, 6.9, 6.10, 6.12, 6.15 and 6.16 Polylogic Ltd.

Chapter 7
7 Chiron World Sports Cars Ltd; 7.1 Josh Smith; 7.3 Pro-Bolt Ltd; 7.5 Josh Smith (T) and Pro-Bolt Ltd (B); 7.7 Ben Lowden; 7.9 Demon Tweaks except 'gold foil' Gustoimages/Science Photo Library; 7.10 and 7.11 Josh Smith; 7.13 Radical Sports Cars.

Chapter 8
8 Will Belcher Photography; 8.1 Reproduced with the kind permission of the MSA; 8.2 Alastair Staley/GP2 Series Media Service/LAT Photographic; 8.6 helmet- Josh Smith, HANS device- HANS, protective suit- Alpinestars S.p.a.; 8.7 MK Technologies Ltd; 8.8 P.A.C.E. Performance and Competition Engineering; 8.9 exe-tc; 8.10 www.demon-tweeks.co.uk; 8.11 MK Technologies Ltd; 8.12 Chiron World Sports Cars Ltd.

Every effort had been made to trace the copyright holders but if any have been inadvertently overlooked the publisher will be pleased to make the necessary arrangements at the first opportunity.